CLASSIC GEOLOGY IN EUROPE 3

Iceland

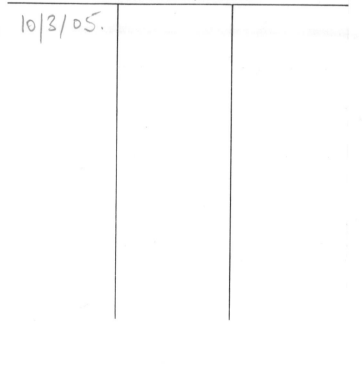

Iceland

Thor Thordarson
University of Hawaii

Armann Hoskuldsson
South Iceland Institute of Natural History

TERRA

First published in 2002 by Terra Publishing

Terra Publishing
PO Box 315, Harpenden, Hertfordshire AL5 2ZD, England
Telephone: +44 (0)1582 762413
Fax: +44 (0)870 055 8105
Website: http://www.terrapublishing.co.uk
E-mail: publishing@rjpc.demon.co.uk

ISBN: 1-903544-06-8 paperback

12 11 10 09 08 07 06 05 04 03 02
10 9 8 7 6 5 4 3 2 1

British Library Cataloguing-in-Publication Data
A CIP record for this book is available from the British Library

Library of Congress Cataloging-in-Publication Data are available

Typeset in Palatino and Helvetica
Printed and bound by Biddles Ltd, Guildford and King's Lynn, England

Contents

Preface

Iceland is a geological marvel where the elements of nature appear to spring to life right in front of you. From a personal experience, we have no hesitation in declaring it to be one of the most dramatic "geowonderlands" on Earth, in a class with Hawaii and New Zealand. Mother Earth is very much alive in Iceland, not only continually reshaping the land but also the cultural awareness of the inhabitants. As such, it is an integral part of the Icelandic psyche and an essential element in obtaining a full appreciation of the country's history. Thus, the purpose of this book is to present some basic information on the geology of Iceland and to provide some insights on its influence in shaping Icelandic culture.

In writing a book such as this, authors rely on innumerable written sources representing countless hours of work by a large group of researchers. However, it is impractical to include hundreds of direct citations to individual works in a text of this nature, as it would make it very difficult for the introductory reader. Nevertheless, we would like to extend our gratitude to the scientists whose work has made the writing of this book possible, and we dedicate it to them. Many of these contributions are listed in the Bibliography, and acknowledgements for illustrations borrowed with permission from other texts are listed overleaf.

We also like to extend thanks to Alwyn Scarth and Bruce F. Houghton, who reviewed parts of the text and made many valuable suggestions; to Carol Whitney Thordarson, for her laborious editing of early versions of the text and for her understanding and patience during the time of writing, which inevitably eroded valuable family time on many evenings and weekends. Finally, we extend our gratitude to Roger Jones for all his support, patience and encouragement.

Thor Thordarson
Armann Hoskuldsson
July 2002

Acknowledgements

The authors would like to acknowledge the following sources for permission to use copyright material in the preparation of the illustrations, all of which have been modified. The full bibliographic details are in the Bibliography (pp. 188–190). All other illustrations are the work of the authors themselves.

Figure	Source	Figure	Source
1.1	Saunders et al. (1997)	5.2	Steinþórsson (1966)
1.2	Sæmundsson (1979)	5.4a	Ray Miller
1.3a	Jóhannesson (1980)	5.5	Larsen (2000)
1.3b	Steinþórsson (1981)	5.7a,b	Thordarson & Self (2001)
1.4	Steinþórsson (1981)	5.8	Thordarson & Self (2001)
1.5	Sæmundsson (1979)	5.10	Thorarinsson (1958)
1.6	Gudmundsson (1995)	5.11	Thorarinsson (1958)
1.7a	Alfreð Jónasson, courtesy of Sólarfilma	6.1a	Jóhannesson & Sæmundsson (1998)
1.7c	Marta Bergman & Bergsteinn Gizurarson	6.1b	Walker (1963)
1.8c	Thordarson & Self (1998)	6.2	Jónsson (1988)
1.9b	Lockwood & Lipman (1980)	6.4	Blake (1966)
1.9d	McPhie et al. (1993)	6.5	Walker (1963)
1.13a	John Sinton	6.6b	Walker (1966)
1.14	Sæmundsson (1979)	6.8a,b	Walker (1962)
1.15a	Einarsson (1991)	7.1	Jóhannesson & Sæmundsson (1998)
1.19	Geirsdottir & Eiriksson (1994a)	7.4	Einarsson (1991)
1.20	Walker (1966)	7.6	Sæmundsson (1991)
2.1	Jóhannesson & Sæmundsson (1998)	7.8a	Sæmundsson (1991)
2.2	Geirsdottir & Eiriksson (1994b)	7.9	Sæmundsson (1991)
2.3	Geirsdottir & Eiriksson (1994b)	7.10	Sæmundsson (1991)
2.6	Kjartansson (1973)	8.1	Jóhannesson & Sæmundsson (1998)
2.7b	Sigurgeirsson (1995)	8.2	Norddahl & Haflidason (1992)
3.1a	Jóhannesson & Sæmundsson (1998)	9.1	Jóhannesson & Sæmundsson (1998)
3.1b	Einarsson & Bjornsson (1979)	9.2	Jóhannesson (1980)
3.2	Einarsson (1991)	9.3	Bjórn Rúnilsson
3.3	Sæmundsson (1992)	9.4	Jóhannesson (1980)
3.4b	Einarsson (1991)	9.5	Einarsson (1991)
3.7	Larsen & Thorarinsson (1977) and Larsen et al. (1999)	10.4	Gunnarsson (1987)
4.1	Jakobsson (1968)	10.5	Larsen (1984)
5.1	Jóhannesson & Sæmundsson (1998)		

Introduction

Iceland is one of the very few places where we can directly observe the Earth's growth processes in action. We all are aware of two geological fundamentals: the tangible materials that we walk on and the sometimes intangible processes that operate on our surroundings. The materials (excluding living organisms) are the rocks that make up the solid crust or the outermost shell of the Earth, and the processes represent the forces that construct and erode these crustal rocks and thus shape the landscape that we have before us. The processes that operate at or near the Earth's surface are generally categorized as external, whereas those that are at work deep below the surface, or result from such activities, are called internal.

Iceland is an exceptional natural laboratory where almost all of the principal geological processes can be observed and it is an ideal setting for laymen, as well as professionals, to witness how Mother Nature operates. Unusually high rates of volcanic activity and dynamic fault movements, coupled with rapid erosion and efficient transport and deposition of sediment, make Iceland a diverse environment where constructive processes outweigh destructive ones. Nowhere on Earth is the architecture of the spreading sea floor better exposed, where different patterns of fault lines and fault movements, along with chains of volcanoes, characterize the parts of Iceland interpreted as the boundary between the Eurasian and American plates.

Large icecaps and extensive river systems rapidly grind down the volcanic pile, dispersing vast amounts of debris from the highlands to the lowlands to form thick sequences of glacial, fluvial and lacustrine sediments. Along the seashore these formations are further modified by the pounding waves of the North Atlantic, where the often spectacular landforms exhibit an intricate balance between land construction and erosion.

Over the past 7 million years Iceland has been situated at a boundary of major air and ocean masses, and consequently has been exposed to extreme climatic changes. The rocks in Iceland clearly reflect this fluctuating climate. The dynamic interplay between constructive and destructive geological processes has resulted in the juxtaposition of volcanic and sedimentary rocks. Dramatic swings in the climate over the past 7 million years are clearly illustrated by the occurrence of diverse rock types in alternating volcanic and sedimentary successions. These changes have also

1

Figure 0.1 Hekla in South Iceland, the queen of Icelandic volcanoes, viewed from southwest.

resulted in a wide range of volcanic structures, because the eruptions occurred in environments ranging from subaqueous through subaerial to subglacial (Fig. 0.1). Consequently, each rock formation reflects the flavour particular to the builders – the spectrum of internal and external processes that moulded Iceland into its current shape.

Iceland is the youngest part of a much larger volcanic province that extends from Scotland to Greenland. It is also the only part of the province that is still active and possibly the only place on Earth where the processes involved in the construction of this type of volcanic province can be observed directly. Furthermore, it provides a window into continent formation as it was early in Earth's history, because such continents are likely to have been produced by volcanism of a nature similar to that found in Iceland.

There are various ways one could organize a book of this kind, but we have chosen a format that follows an excursion that a visitor could conveniently take around Iceland. This gives us the flexibility to organize the book as a set of trips that are arranged in a rational order from the traveller's point of view. It also allows us to focus our attention on regions and localities that are popular among those who visit Iceland, as well as to point out interesting geological localities of which the reader might otherwise be unaware. Consequently, the book is by no means a comprehensive textbook on Icelandic geology, but is intended for those who want to acquire some basic information on a subject of a particular geological phenomenon. We have chosen to eliminate bibliographic references, as they would have made the text too cluttered, but a selection of further reading is provided at the end. We have also tried to restrict the use of technical terms to a minimum, but those that are necessary are explained in the Glossary and are highlighted in **bold type** on first appearance.

2

The first chapter provides a brief overview of the geology of Iceland and the forces that have shaped the island through its relatively short geological existence. It also includes a brief explanation of the terminology used throughout the book. In chapters 2–9 we present a set of geological excursions covering the most accessible parts of Iceland, the Lowlands. In general, the excursions are planned such that they follow or extend from Highway 1, the main artery around Iceland. We begin our excursions in Chapter 2 at the Reykjanes Peninsula, because it is the point of arrival for most travellers. The excursions in subsequent chapters are arranged in a counterclockwise fashion around the periphery of the island. The breakdown of excursions into chapters generally follows the traditional geographical subdivisions, similar to that of the 1:250 000 topographical and geological maps of Iceland. In Chapter 10 we finish up with a selection of excursions within the Interior Highlands

The heavy emphasis on the regions closest to Reykjavík is deliberate, because many travellers spend only a few days in Iceland. Consequently, they cannot travel all around the country, but can easily make short 1–2 day trips out of Reykjavík covering the area from Snæfellsnes to Öræfajökull. However, we believe that we have included enough details from other regions to provide a geological guide to all of Iceland.

Those who are interested in more information on the geology of Iceland are encouraged to look up the references listed in the Bibliography. Geological maps of Iceland are produced by the Iceland Geodetic Survey and Museum of Natural History, and can be obtained in local bookstores. Finally, all Icelandic names will appear in Icelandic (i.e. using the Icelandic alphabet), so that the reader will be able to relate the place names mentioned to maps and road signs.

3

Travel in Iceland

We would like to offer a few words of warning in respect of travel arrangements. Despite Iceland's location just south of the Arctic Circle, it has a relatively mild maritime climate, with a 4-month-long growing season and typical July temperatures around 10–15°C. However, the northerly position of Iceland puts it at the boundary between warmer temperate and colder arctic air masses, and therefore the weather can change drastically and very quickly. It is often said that one can experience all four seasons in the space of a day and we know from experience that this is true. Thus, proper gear is essential, as well as a healthy respect for the elements.

In Iceland the flora and fauna manage to survive at this far edge of the inhabitable environment. Consequently, vegetation is extremely fragile and easily ruined by human intervention. This is especially true of the vegetation up in the Central Highlands. In this regard we encourage responsible behaviour. Please respect the guidelines for travel provided by the authorities and use only designated paths and tracks when travelling off the main roads.

Iceland can be considered an open country because access to explore its natural phenomena is normally not restricted, but respect for private property and open land is expected. Note also that potentially hazardous areas, such as cliffs, waterfalls, volcanic craters and hot springs, are often not roped or fenced off, so be careful where you step. Furthermore, fording a stream, even in a four-wheel-drive vehicle, can prove extremely hazardous because many streams and rivers are fed by glacial meltwater. Their discharge can fluctuate wildly during the course of a single summer day – water levels being at their lowest in early morning and increasing throughout the day. As you explore and enjoy the natural wonders of Iceland, respect the elements and the environment, and remember that individuals are responsible for their own safety. As with any outdoor adventure, and especially in Iceland, extreme caution and prudence are advised.[1]

---

1 For further information on these matters, contact the Icelandic Tourist Bureau (Ferðamálaráð Íslands) at Lækjargata 3, Gimli, IS-101 Reykjavík (tel.: +354-535-5500, fax: +354-535-5501) or at http://www.icetourist.is.

Chapter 1

The geology of Iceland

Introduction

Iceland is located in the North Atlantic Ocean between Greenland and
Norway at 63°23′N to 66°30′N. It is a landmass that is part of a much larger
entity situated at the junction of two large submarine physiographic structures, the Mid-Atlantic Ridge and the Greenland–Iceland–Faeroes Ridge
(Fig. 1.1). As such, Iceland is a part of the oceanic crust forming the floor
of the Atlantic Ocean. This region is known as the Iceland Basalt Plateau
which rises more than 3000 m above the surrounding sea floor and covers
about 350 000 km². About 30 per cent of this area (103 000 km²) is above sea
level, the remainder forming the 50–200 km-wide shelf around the island,
sloping gently to depths of about 400 m before cascading into the abyss.

Iceland is geologically very young and all of its rocks were formed
within the past 25 million years. The stratigraphical succession of Iceland
spans two geological periods: Tertiary and Quaternary (Table 1.1). The
construction of the Iceland is thought to have begun about 24 million years
ago, but the oldest rocks exposed at the surface are 14–16 million years old.
If we take the age of the Earth as one year, then Iceland was only born less
than two days ago. The first regional glaciers of the Ice Age appeared in
Iceland about five hours ago and only a minute has passed since the
Holocene warming removed this ice cover from Iceland (Table 1.1).

The surface of Iceland has changed radically during its brief existence.
The forces of nature that constantly mould and shape the face of the Earth
operate faster in Iceland than in most other places. The rocks are shattered
by the frequent change of frost to thaw, and the wind, seas and glaciers
laboriously grind down the land. Erosion removes about a million cubic
metres of land from Iceland each year, but volcanism and sedimentation
more than counterbalance this loss. The latter processes truly have the
upper hand in this continuing battle, as is evident in the landmass that is
now Iceland.

Figure 1.1 Iceland is an elevated plateau in the middle of the North Atlantic, situated at the junction between the Reykjanes and Kolbeinsey Ridge segments. The dashed line shows the position of the Iceland mantle plume from 65 million years to the present day.

Geological setting: a global perspective

The prologue to the formation of Iceland includes the continental break-up that separated Newfoundland and Greenland from Europe, and the subsequent formation of the sea floor that now surrounds Iceland (Fig. 1.1). About 400 million years ago the continents on either side of the North Atlantic Ocean had merged to form a continuous landmass, which remained intact for more than 300 million years. About 70 million years ago this large continent began to break up along a fracture zone extending from the latitudes of Newfoundland–British Isles northwards into the Arctic between Scandinavia and Greenland. As the continent on either side of the fracture drifted apart, a new plate boundary was formed.

This plate boundary is where active spreading and plate growth take

6

Table 1.1 Geological timetable for Iceland, with terminology used in the text; ages in years and millions of years (my).

Era	Period	Epoch	Age	Stage	Substage	Formations	Events
Cenozoic	Quaternary	Holocene	0	Sub-Atlantic			
			2500	Sub-Boreal			
			5000				
			7200	Atlantic			
			9300	Boreal			
			10000	Pre-Boreal			
		Upper Pleistocene	11000		Younger Dryas	Upper Pleistocene	Ice Age glacier disappears
			12000	Weichselian	Allerød		Fossvogur sediments accumulate
			20000		Older Dryas		
			70000				Last glacial stage
			130000	Eemian			Elliðavogur sediments accumulate Last interglacial
			170000	Saale			Penultimate glacial stage
			700000				Svinafell sediments accumulate
		Lower Pleistocene				Plio-Pleistocene	Rauðsgjá tillite
			2.5 my				Breiðavik tillite and sediments Fufuvík tillite formed Full-scale glaciation
							Tjörnes sediments stop accumulating
	Tertiary	Pliocene	3.3 my				Pacific Ocean fauna arrive in Iceland Bering Strait opens
							Tjörnes sediments begin to form
			7 my				First sign of cooling climate
		Upper Miocene	12 my			Tertiary Basalt	Warm temperate climate
		Middle Miocene	18 my				Oldest rock on land
		Early Miocene	25 my				Birth of Iceland

place. It can be viewed as a suture where the crust is being pulled apart and molten rock (i.e. magma) wells up from below to fill in the gash. As spreading continued, more magma welled up to seal the cracks and thus gradually produced the crust that forms the ocean floor around Iceland. The spreading of the sea floor is a piecemeal process and in the North Atlantic

7

the spreading rate is about 2 cm/year (i.e. 1 cm/year in each direction). The plate boundary is delineated by series of faults and volcanoes, which together form a distinguishing ridge-like structure in the middle of the ocean. This structure is known as the Mid-Atlantic Ridge and its crest delineates the spreading axis. It is the suture where the American plate (to the west) and the Eurasian plate (to the east) are actively being pulled apart by the forces of plate motions (Fig. 1.1). Such mid-ocean ridges are found in all of the major oceans on Earth.

Normally, the mid-ocean ridges do not build up above sea level, but they do in Iceland. More eruptions than usual occur on the plate boundary across Iceland because there is a hotspot under the ridge. The hotspot is a kind of plume of hotter and less dense material within the Earth's mantle. It is more buoyant than its surroundings and thus rises towards the surface. This structure is usually referred to as the Iceland mantle plume and it has been active for the past 65 million years. During this period it has brought unusual amounts of magma to the surface. As a result, a series of volcanic regions formed, which stretch across the Atlantic from Scotland to Greenland, including Iceland, which is the youngest and the only one that is still active. Collectively these volcanic regions are known as the North Atlantic Large Igneous Province. It is approximately 2000 km long and it represents about 10 million cubic kilometres of magma that has emerged from the Iceland mantle plume through volcanic activity over 65 million years. This figure is roughly 50 times the volume of Iceland.

Geological framework

Iceland is located at the junction between the Reykjanes Ridge in the south and the Kolbeinsey Ridge in the north. The ridges represent submarine segments of the mid-ocean ridge (i.e. the plate boundary) that is closest to Iceland. The surface expression of the plate boundary in Iceland is the narrow belts of active faulting and volcanism extending from Reykjanes in the southwest, which zigzag across Iceland before plunging back into the depths of the Arctic Ocean of Öxarfjörður in the north (Fig. 1.2). This plate boundary is Iceland's major geological showpiece because it is the only section of the Mid Atlantic Ridge exposed above sea level.

Above the plate boundary, the spreading rips apart the brittle crust and results in the formation of wide cracks and faults, both of which are orientated perpendicular to the spreading directions, which in Iceland are 105°E and 285°W. The spreading also results in the formation of vertical dykes that sometimes become pathways for the magma to the surface. Above ground, these rifts appear as swarms of linear volcanic fissures

Figure 1.2 The principal elements of the geology in Iceland, outlining the distribution of the major geological subdivisions, including the main fault structures and volcanic zones and belts. RR, Reykjanes Ridge; RVB, Reykjanes Volcanic Belt; WVZ, West Volcanic Zone; MIB, Mid-Iceland Belt; EVZ, East Volcanic Zone; NVZ, North Volcanic Zone; TFZ, Tjörnes Fracture Zone; KR, Kolbeinsey Ridge (KR); ÖVB, Öræfi Volcanic Belt; and SVB, Snæfellsnes Volcanic Belt.

confined to narrow belts know as the volcanic zones (Fig. 1.2). The volcanic zones are connected by large **transform faults** known as **fracture zones** or, when volcanically active, as **volcanic belts**. Together, these structures cover about one third of Iceland (30 000 km³); the nomenclature used for these structural identities is given in Figure 1.2.

It is important to recognize that the volcanic zones are 20–50 km-wide belts and their magma production more than matches the plate movements. Consequently, the magma that emerges at the surface through volcanic activity accumulates within the volcanic zones, more so towards the centre than to the margins. Thus, the volcanic successions in the centre of the rifts are buried rapidly and follow a steep path as they move away from the spreading centre (Fig. 1.3). However, the successions closer to the margins of the volcanic zones keep to much shallower paths as they travel away from the spreading axis. The excess load in the middle of the zones

9

also causes downsagging of the crust, such that the successions closer to the margins are tilted, acquiring a shallow dip (5–10°) towards the spreading axis. This tilt remains with the rock pile as it drifts out of the volcanic zones and is accreted to the plates on either side. Thus, together, the spreading and the volcanism produce a shallow **syncline**, which explains why the regional dip of the flanking older successions is generally towards the currently active volcanic zones (Figs 1.2, 1.3).

As mentioned earlier, the construction of Iceland resulted from interaction between a spreading plate boundary and a more deeply rooted mantle plume. The overall morphology and geological architecture of Iceland is a representation of this interaction and its development through time. The most obvious manifestation of the interaction is the elevation of Iceland over the surrounding sea floor. Here the buoyant mantle plume has pushed up the crust to form a large oval bulge. The submarine Faeroe–Iceland and Greenland–Iceland ridges on either side of Iceland are older traces of this bulge that have been displaced from their original position by seafloor spreading.

The centre of the Icelandic mantle plume is below the northwestern part of Vatnajökull icecap (Fig. 1.1). It is modelled as a 200–300 km-wide cylindrical zone of highly viscous semi-solid material that is hot and buoyant, rising extremely slowly from depths of 400–700 km. As it approaches the surface, parts of the plume melt and thus provide magma to the volcanoes. About 24 million years ago the centre of the mantle plume was positioned at the extrapolated intersection of the Reykjanes Ridge and the Kolbeinsey Ridge.

Although the plate-tectonic model outlined above explains the gross features of the regional geology of Iceland, it does not fully conform to all established geological facts. Distinctive reversals in dip directions in Borgarfjörður, Vatnsnes and Hreppar areas indicate broad anticlines that follow the same axial trends as the rifts. These anticlines are also associated with major unconformities and breakdown in the trends of isochrons (lines of equal age) across Iceland (Fig. 1.3). These discrepancies occur because the relative positions of the spreading axis and the mantle plume have changed with time.

If we assume that the mantle plume is a stationary structure, then the position of the spreading axis must have changed by migrating towards the west-northwest at 0.3 cm/year. This movement is in addition to the actual spreading motion that separates the plates and it occurs concurrently. As the active spreading axis moves away, the plume responds by re-adjusting the position of the axis and forming new rifts closer to its centre. The rifts farther away gradually become inactive (Fig. 1.4). This process is called a rift jump.

Figure 1.3 A stylized cross section showing the general structure of the Icelandic crust from the Snaefellsnes Peninsula across the West Volcanic Zone. Loading by volcanism tilts the strata towards the volcanic zones, forming a shallow syncline centred on the spreading axes and a shallow anticline in the region between the volcanic zones. The inset graph shows the spreading paths for crustal elements (black rectangles) formed within different parts of an actively spreading volcanic zone. Rocks formed near the edge of the zone follow a shallow spreading path, whereas those formed close to the spreading axis follow a much steeper path. The dashed lines are planes of equal age and the dotted lines are isotherms. The shaded area represents a region where the temperatures are high enough to begin to melt the subsiding rock masses.

mantle plume centre

Figure 1.4 Cartoon of structural developments on a regional scale as a result of rift jumps caused by westward migration of rift zones relative to stationary mantle plume.

According to this model, the currently active rifts are the West and North Volcanic Zones. The East Volcanic Zone is a rift in the making that eventually will take over from the West Volcanic Zone (Fig. 1.2). Similarly, the precursor rift to the current West and North Volcanic Zones is a broad syncline exposed in the Tertiary succession of Snæfellsnes in West Iceland and Víðidalur in North Iceland. The key consequences of rift jumps are the trapping of older crust between the two rift segments and the overlap in activity on the established versus the developing rift segments with respect to time. The accumulation of volcanic material within the rift zones results in sagging of the crust, and in tilting of the adjacent rock piles towards the volcanic zones. Consequently, the trapped crustal segment acquires the form of a broad anticline (Fig. 1.3). Thus, shifts in the position of the spreading axis explain the occurrence of synclines and anticlines in the geological succession of Iceland.

There are two active **intraplate volcanic belts** in Iceland where young (<2 million years old) volcanic rocks rest unconformably on older formations, indicating a significant interval in volcanic activity and outpouring of magma. The first is the Öræfi Volcanic Belt, to the east of the plume centre and the current plate margins. Second is the Snæfellsnes Volcanic Belt in West Iceland, which sits on the mantle-plume trail and is in part superimposed on an extinct volcanic zone, precursor to the West Volcanic Zone.

VOLCANISM

The Öræfi Volcanic Belt probably represents an embryonic rift. It is possible that the westward migration of the entire Iceland rift system is forcing the plume to establish a new volcanic zone by melting its way through the crust of East Iceland. Thus, the Öræfi Volcanic Belt may indicate yet another jump in the spreading axis across Iceland.

Volcanism

Iceland is well known for its volcanic activity. Questions often asked are:
• How many eruptions have taken place in Iceland?
• What is the number of active volcanoes?
The answers are not as straightforward as might be expected. The definitions of a volcano and an eruption encompass a wide range of phenomena that deviate significantly from the simple image of a conical mountain with smoke rising from its top.

It is not easy to say how many eruptions have occurred in Iceland throughout its geological history because it is not always obvious what part of the rock sequence represents a single eruption. However, recent volcanic activity in Iceland shows that on average there is an eruption every five years. This translates into about 200 eruptions in the past 1000 years. Using this frequency as a guide, the total number of eruptions that may have taken place since the birth of Iceland, 24 million years ago, is in the order of 5 million.

The answer to the second question is even more difficult because it is not clear how many of the volcanoes in Iceland should be considered active. Although it is easy to count the central volcanoes that have erupted time after time in the recent geological past (20 in total; Fig. 1.5), the difficulty arises when we try to assess the status of volcanoes that erupted only once. By definition, such volcanoes are extinct, but the region in which the volcanoes are situated may still be volcanically active. To circumvent this problem, geologists have introduced the concept of a volcanic system.

A volcanic system is the principal geological structure in Iceland: a fissure swarm or a central volcano, or both, which are surface expressions of two different types of subsurface magma-holding structures, the first a deep-seated magma reservoir, the second a shallower crustal magma chamber (Fig. 1.6). Each volcanic system is characterized by conspicuous tectonic architecture and distinct magma chemistry, and typically has a lifetime of 0.5–1.5 million years. Altogether there are 31 active volcanic systems in Iceland (Fig. 1.5).

The fissure swarms are narrow and elongated strips (5–20 km wide and 50–100 km long) of tensional cracks, normal faults and volcanic fissures

13

Tertiary Basalt formation

Plio-Pleistocene formation

Upper Pleistocene formation

Volcanic system

─ fissure swarm

─ central volcano

summit crater or caldera

1. Reykjanes	12. Tungnafellsjökull	23. Fremrinámur
2. Krýsuvík	13. Vestmannaeyjar	24. Krafla
3. Brennisteinsfjöll	14. Eyjafjallajökull	25. Þeystareykir
4. Hengill	15. Katla	26. Öræfajökull
5. Hróðmundartindur	16. Tindfjöll	27. Esjufjöll
6. Grímsnes	17. Hekla	28. Snæfell
7. Hrafnabjörg	18. Torfajökull	29. Ljósufjöll
8. Prestahnjúkur	19. Veiðivötn	30. Helgrindur
9. Kjölur	20. Grímsvötn	31. Snæfellsjökull
10. Hofsjökull	21. Kverkfjöll	
11. Kerlingarfjöll	22. Askja	

Figure 1.5 Distribution and components of active volcanic systems in Iceland.

(Fig. 1.6). They are the surface manifestations of elongated magma reservoirs at the base of the crust (>10 km depth). These swarms are typically aligned subparallel to the axis of their host rift zone, illustrating that the fundamental force responsible for their formation is plate spreading. Wide cracks indicative of pure crustal extension are usually the most conspicuous structures on the surface. Fault scarps and **graben** are also common and indicate vertical displacement and extension of crustal blocks. Young volcanic fissures typically appear as a row of small volcanic cones, whereas fissures that erupted beneath the Ice Age glaciers occur as elongated **móberg** ridges.

When present within a volcanic system, the central volcano is the focal point of eruptive activity. It is the surface manifestation of the crustal magma chambers that lie 2–6 km below the surface. Usually, the central volcanoes are the largest volcanic edifice within each system and are often capped by a collapsed crater known as a caldera (Fig. 1.6).

14

Figure 1.6 The main structural elements of a volcanic system. The numbers indicate the growth sequence of a dyke rising through the crust in a major eruption episode.

Events on the volcanic systems are intimately linked to the plate movements. The spreading and subsequent rifting of the crust that take place at the plate boundary are not continuous in either time or space. They occur as distinct rifting episodes confined to a volcanic system at any one time. Normally, the whole system is activated in these episodes, which typically last between several years and decades. Recurring earthquake swarms and volcanic eruptions on the fissure swarm and within the central volcano characterize these rifting episodes. It is traditional in Iceland to refer to such eruptions as eldar, which translates as "fires" (e.g. the Krafla Fires).

Magma types
Magmas are fluids that are formed by melting of rocks in the upper mantle or the lower crust at 700–1300°C. Melting at high temperatures (>1100°C) produces mafic magmas that are relatively low in silica (<52%) and rich in magnesium, iron and calcium. On the other hand, felsic magmas are formed by melting at relatively low temperatures (700–900°C) and are rich in silica (>64%), but also contain significant amounts of aluminium, sodium and potassium (Table 1.2). Magmas of intermediate compositions (<52–64% SiO$_2$) are formed by melting at intermediate temperatures or by mixing of mafic and felsic magmas.

Because temperature increases steadily with depth below the surface, mafic magmas generally formed at greater depths than felsic magmas.

15

Table 1.2 Classification of magma types, based on chemical composition and nomenclature of igneous rocks, with reference to environment of solidification.

Environment of emplacement	Geological terms		Mafic (<52% SiO$_2$)	Intermediate (52–64% SiO$_2$)	Felsic (>64% SiO$_2$)
Surface	Extrusive	Volcanic	Basalt	Andesite	Dacite/ rhyolite
Upper crust	Intrusive	Hypabyssal	Dolerite (diabase)	Microdiorite	Microgranite
Lower crust	Intrusive	Plutonic	Gabbro	Diorite	Granite (granophyre)
Colour			Dark	Intermediate	Light
Colour index			Melanocratic	Mesocratic	Leucocratic

Consequently, the deeper-seated magma reservoirs (Fig. 1.6) contain mafic magma that is formed by **partial melting** of the mantle. Magma rising from the reservoirs is the main source of volcanic eruptions on the fissure swarms, which accounts for the dominance of basaltic volcanism there.

The story is very different for the magma chambers located in the crust below the central volcanoes, because they may contain all three magma types. Mafic magma accumulates in the magma chambers from injection below, whereas felsic magma is produced by partial melting of the sur-rounding crustal rocks. Mixing of the other two magma types forms inter-mediate magmas. Magma chambers are the source of eruptions at central volcanoes, which explains why they are the only volcanoes in Iceland that erupt intermediate and silicic magmas in addition to basaltic magma.

The magma rises from depth towards the surface through conduits contained within the volcanic systems. In the process it cools and crystal-lizes to form igneous rocks. Magmas that are injected into the Earth's crust, and cool and crystallize below the surface, are called intrusive rocks. Those that become trapped in the deeper and hotter part of the crust cool slowly to form coarsely crystalline bodies, known as plutonic rocks. In Ice-land such rocks occur in the root zone of central volcanoes and represent solidified magma chambers. They are grouped according to composition into gabbro, diorite and granite.

Closer to the surface the temperature is lower and the magma cools a little faster, resulting in the formation of medium to fine-grain hypabyssal rocks. Intrusive rocks of this type typically occur as sills and dykes, and are named dolerite, microdiorite, and microgranite in accordance with their chemical make up. Sills are aligned horizontally or parallel to the strati-fication of the surrounding rocks. Dykes are vertical structures that cut across the stratification and often represent feeder conduits to eruptions.

Magma erupted onto the Earth's surface cools very rapidly to form fine-grain to very fine-grain extrusive or volcanic rocks, grouped as basalt, andesite and rhyolite, in accordance with their chemical composition.

Rocks

As different types of cells unite in a variety of combinations to form different organisms, **minerals** combine in innumerable ways to form a perplexing variety of rock types. Consequently, it is the mineral types and their relative abundance that determine the characteristic texture and the bulk chemical composition of a rock, just as the network of cells controls the shape and make-up of living organisms. In the same way as we divide living organisms into the animal and plant kingdoms, rocks can be grouped on the basis of their origin and overall characteristics into three families: **igneous, sedimentary**, and **metamorphic**. Igneous (fire-formed) rocks are formed by solidification of magma as the result of cooling, whereas sedimentary (settling) rocks are formed by induration of sediment by cementation or other processes at or near the Earth's surface. Metamorphic ("change of form") rocks are formed by re-crystallization of pre-existing solidified igneous or sedimentary rocks as they are buried in the Earth's crust and are re-introduced to high temperatures or high pressures, or both. Metamorphic rocks are absent from Iceland's stratigraphical succession, but are found as ice-rafted boulders on beaches.

The Earth's crust was initially formed from magma and about 95 per cent of it is igneous rocks or their metamorphosed equivalent. Sedimentary rocks are confined to the outermost skin of the crust (i.e. the top 2–3 km). On the continents about 75 per cent of the rocks we see are sedimentary, mainly because the external processes that operate relentlessly to produce them are confined to the Earth's surface. The internal processes are mostly hidden from our view and so is the bulk of the igneous and metamorphic rocks that they produce.

In Iceland the situation is reversed, because the majority (≥75%) of the exposed rocks are igneous and as such they reflect the **lithologies** that make up the Earth's crust. About 75 per cent of the igneous rocks in Iceland are mafic; the rest are intermediate or felsic. The other 25 per cent of the exposed rocks are sedimentary. Metamorphic rocks such as slate, schist and gneiss are not present.

Volcanic eruptions and their products

A volcanic eruption refers to expulsion of magma, gas or rocks, or all three, in any shape or form, onto the Earth's surface. Consequently, a volcanic eruption can be a single explosion that lasts only a few seconds or it can represent a continuous effusion of lava that lasts for years or even decades. An eruption can also feature a single style of activity, either purely explosive or effusive, or it can encompass both styles, because many eruptions that begin explosively transform into quiet effusion of lava over time. As a result, the definition of a volcanic eruption is not simple. It is customary

Figure 1.7 (a) Effusive eruption, (b) explosive eruption, (c) mixed eruption.

among volcanologists to take any near-continuous expulsion of any combination of magma, gas and rocks, whether they are erupted in a single pulse or continually over period of time, to be a volcanic eruption.

In essence, there are two fundamental styles of eruptions, effusive and explosive (Fig. 1.7). Effusive eruptions are characterized by a more quiet extrusion of magma, which flows away from the vent(s) as a coherent body of lava. In explosive eruptions the magma is disintegrated by the rapid expansion of gases or steam, producing rock fragments that are collectively known as tephra. Many eruptions feature both styles of activity, producing both tephra and lava, and are referred to as mixed eruptions.

Effusive eruptions and types of lava flows   Effusive eruptions occur when magma pours out onto the surface as lava flows. Lava flows occur in various sizes and shapes, and their form is a good indicator of how readily they flowed. The fluidity of lava is controlled by its viscosity, which in turn is determined by its chemical make-up. Therefore, on the basis of magma composition, effusive eruptions are divided into three categories: mafic, intermediate and felsic. At one end there are fluid mafic magmas that produce thin sheets of lava, often covering large areas, and at the other are highly viscous felsic magmas that form thick stubby flows.

Pahoehoe lavas (helluhraun in Icelandic) are formed on land by outpourings of very fluid mafic magma in effusive eruptions. Pillow lavas (bólstraberg in Icelandic) are formed when such eruptions occur under water. Both lava types refer to **compound lavas**, consisting of many round lobes with smooth outer surfaces.

Pahoehoe lava flows vary greatly in size, with thicknesses ranging from a few metres to several hundred metres, and lengths from 1 km to more than 120 km. Each flow is typically made up of countless lava lobes that range in size from toe-like lobes a few centimetres thick (Fig. 1.8a) to sheet-like lobes tens of metres thick.

Similarly, pillow lavas consist of many bulbous sack- or pillow-shape lobes that in long section are often tube like (Fig. 1.8b). The space between pillow lobes is often filled with ash-size debris consisting of angular glass fragments formed by disintegration of the pillow surface as a result of quenching in water.

In pahoehoe and pillow, the lava is transported from the vents in lava tubes beneath a coherent stationary crust. The crust is highly insulating and it minimizes cooling of the lavas during transport to the active flow front. As the lava breaks out of the tube, it forms a new lobe that is immediately sealed with new crust. The lobe grows in thickness and length by continuous injection of lava into its molten interior. Eventually, cooling prevents further lobe growth and the flow of lava is entirely confined to

VOLCANISM

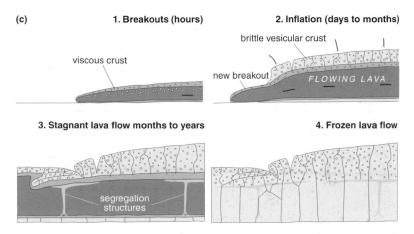

(c)   1. Breakouts (hours)   2. Inflation (days to months)

brittle vesicular crust

viscous crust

new breakout   FLOWING LAVA

3. Stagnant lava flow months to years   4. Frozen lava flow

segregation structures

Figure 1.8 (a, opposite) Pahoehoe lava. **(b, opposite)** Pillow lava. **(c)** Snapshots showing growth of a pahoehoe lobe by inflation from initial breakout at the active flow front (1) to last stages of lava emplacement (3). Arrangements of internal flow structures in a solidified lava lobe are shown in (4).

the new extension of the tube in the core of the lobe. A new lobe is formed at the front of the extended tube and the process described above is repeated. This mode of emplacement is known as lava inflation (Fig. 1.8c). It is by this step-wise mechanism of forming lobe after lobe that pahoehoe lavas gradually make their way across the landscape without much cooling of the flow interior and in doing so they have formed some of the largest lava flows on Earth.

Aa lavas (apalhraun in Icelandic) are produced by emission of more viscous mafic and intermediate magmas in mixed eruptions. They are generally thicker (3–40 m) and shorter (1–35 km) than pahoehoe flows. Aa lavas are characterized by steep jagged flow fronts and exceedingly rough rubbly surfaces formed by tearing of viscous incandescent lava during flow (Fig. 1.9a). The advance of aa typically occurs in broad open channels, where the lava can flow very quickly (up to 70 km/hr). The advance of the flow fronts is much slower (<2 km/hr) because the lava fans out and slows down as it emerges from the channels. The flow front creeps forwards and steepens until part of it becomes unstable and breaks off. Such collapses happen over and over again as the lava moves slowly forwards over its own rubble. Internally, aa flows consist of coherent lava with highly irregular outlines sandwiched between layers of rubble (Fig. 1.9a).

Block lavas and lava domes are generated by effusive eruption of felsic magmas. They tend to be thick and short lava flows, typically 0.5–3 km long and 40–800 m thick (Fig. 1.9c). The core of these lavas typically consists of flow-banded **stony rhyolite** and is surrounded by a layer of black

21

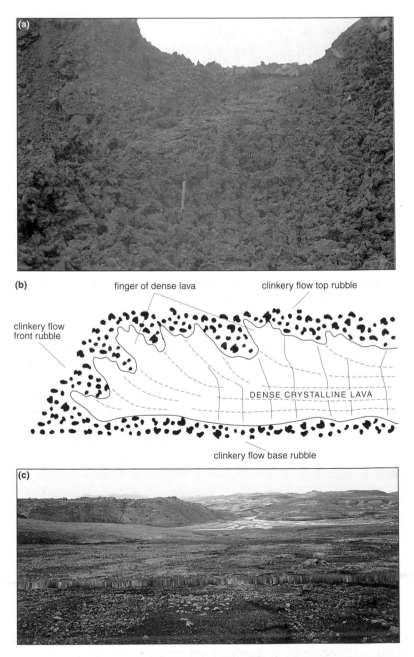

Figure 1.9 Aa and block lava. **(a)** An active aa lava flow; **(b)** a graphic log of aa lava showing key flow structures; **(c)** ~50 m-thick block lava; **(d, opposite)** graphic log showing key flow structures.

The following labels appear on the graphic log (b):

- finger of dense lava
- clinkery flow top rubble
- clinkery flow front rubble
- DENSE CRYSTALLINE LAVA
- clinkery flow base rubble

(d) blocky rubble
vesicular obsidian
obsidian with spherulites

stony rhyolite

vitreous **obsidian**, which may contain **spherulites**. The outermost shell of the lava consists of thick rubble made up of large blocky lava fragments that often reach a diameter of 1 m or more (Fig. 1.9d). When the magma is extruded from the vent, the outer surface cools and solidifies. As a result of continued extrusion of viscous magma into the molten interior, the lava swells, fragmenting the cooler and near-solid lava margins in the process, forming the blocky rubble.

Explosive eruptions and their deposits Explosive eruptions occur when molten magma is fragmented by extremely violent boiling in the upper part of the volcanic conduit. If the explosive fragmentation is solely attributable to the expansion of dissolved gases escaping from the boiling magma, the eruption is said to be magmatic. The power (or explosiveness) of a magmatic eruption is determined by the magma's viscosity and the amount of gas dissolved in the magma at depth. Generally, felsic magmas produce the most powerful explosions because they have the highest viscosity and gas content – a combination that allows for the greatest build-up of internal pressures in the bubbling magma. Mafic eruptions are usually the least explosive because the viscosity and gas content of the magma is low. The bubbles formed expand and escape without significant pressure build-up and the resulting activity is weakly explosive.

If the fragmentation of the magma results from explosive interaction with surface water, groundwater or glacial meltwater, the eruptions are referred to as hydromagmatic. The hydromagmatic explosions are generated by the instantaneous transformation of water into steam upon contact with the much hotter magma. The explosive power of these eruptions primarily depends on the water:magma mass ratio. At optimum mass ratio (~0.3), however, the transformation of water into steam is efficient enough to generate violent explosions that match those of the most powerful magmatic eruptions.

Table 1.3 Grain-size terminology for volcaniclastic deposits and rocks.

Grain size	Volcaniclastic deposits	Volcaniclastic rocks	Other terms often used do indicate specific type of tephra particles	Hydromagmatic clast types
<0.063 mm	Very fine ash	Very fine tuff	Cuspate shards	Blocky shards
0.063–2 mm	Fine–coarse ash	Fine–coarse tuff	Pele's tears Pele's hair	Non-vesicular blocky particles
2–64 mm	Lapilli tephra	Lapilli tuff, lapillistone	Pumice, scoria	Accretionary lapilli, armoured clasts
>64 mm	Bombs (fluidal) blocks (angular)	Agglomerate, volcanic breccia	Spatter, scoria bomb, lithic blocks	Breadcrust bombs, lithic blocks

Tephra or **pyroclasts** are formed when molten magma is fragmented by explosions, and the products of such eruptions are collectively known as **volcaniclastic deposits** or rocks. Volcaniclastic deposits are grouped into categories of ash, lapilli, bombs and blocks. When these deposits are consolidated into a rock they are referred to as tuff, lapillistone, agglomerate or volcanic breccia (Table 1.3).

Pumice and scoria describe certain types of tephra clasts, of lapilli size or larger. Pumice is white to pale grey and has frothy clasts of felsic composition; it commonly floats on water. Scoria, on the other hand, has highly vesicular black to brown clasts of mafic to intermediate composition. Accretionary and armoured lapilli are tephra particles that indicate a hydromagmatic origin when present in volcaniclastic deposits. Accretionary lapilli are bean-like aggregates formed by the coagulation of fine ash in a moist eruption column. Armoured clasts form when a thin layer wet ash coats larger particles such as pumice or scoria. Basaltic tephra deposits often contain drop-like particles called Pele's tears and highly drawn out threads of glass known as Pele's hair; both of these are named after the legendary goddess of Hawaiian volcanoes. In Iceland it is a longstanding local custom to refer to this hair-like volcanic material as witches' hair (nornahár in Icelandic). Blobs of magma that are still molten when they hit the ground and flatten out to form disk-shape particles called spatter.

Explosive eruptions produce a range of **volcaniclastic deposits** and rocks that are categorized into two general groups, flow and **fall deposits**, in accordance with their mode of transport and deposition. Flow deposits are formed by laterally moving currents carrying volcaniclastic debris from a volcano. The most common flow types are pyroclastic surges, pyroclastic flows and lahars, which produce volcanic formations such as base surges, ignimbrites and debris-flow deposits. These currents consist of a dispersive mixture of fluid and particles. The fluid can be gas (volcanic gas, air or steam, or all three) or liquid (water, muddy water or watery mud), whereas the particles can be made up of tephra or freshly fragmented lava or older volcanic rocks, or a mixture of all three components.

Fall deposits typically form widespread but relatively thin tephra layers that drape the land and are divided into scoria, pumice, and ashfall deposits on the basis of dominant fragment type or size. Individual tephra layers tend to become progressively thicker and of coarser grain towards their source volcano. Tephra fall deposits are extremely common in Iceland, whether they occur as distinctive tephra layers in the topsoil or blanket the surroundings of the youngest volcanoes (Fig. 1.10). Some soil profiles contain more than a hundred tephra layers, a record that covers about 7000 years of volcanism. Tephra layers are very useful markers when they occur within stratigraphical successions, because their formation is instantaneous in geological terms and they are widespread. If we know the eruption dates of tephra layers, they can be used to date other geological formations as well as archaeological finds. This method is called tephrochronology and was founded in the mid-twentieth century

Figure 1.10   Historical tephra layers in a soil profile in southern Iceland (source volcanoes and years of eruption).

25

by the Icelandic volcanologist, Sigurður Þórarinsson, through his studies of tephra layers (see Fig. 1.13). Today, tephrochronology is widely used in volcanological studies in Iceland and elsewhere.

Volcanoes – types and architecture
A volcano is any structure produced by volcanic activity, whether it is formed by single or multiple eruptions from the same vent system. At the surface, a volcano consists of two principal components: the vent(s) and the products that issue from it. Although the overall architecture of a volcano is determined by a range of factors, the most significant ones are the type of magma erupted, the habitual eruption behaviour and the shape of the vent system. Consequently, volcanoes display a wide spectrum of shapes, ranging from a hole in the ground (i.e. a crater) to the majestic cone that most people commonly associate with a volcano.

The hole can be either a crater (a roughly conical pit) or a caldera, which is a large circular depression (collapse structure) surrounded near-vertical fault-bounded walls (Fig. 1.11a). The former is created when powerful explosions excavate the ground around the vent(s). The latter is formed when a crustal magma chamber is drained rapidly during an eruption, such that its roof collapses into the void left behind by the departed magma.

In less powerful explosive, mixed or effusive eruptions, some of the erupted material piles up around the vent(s) to form a range of structures such as ramparts, cones and even mountains. The form of these structures depends on whether the volcano was formed by eruption(s) on land, in water, or beneath a glacier, and by the shape of the erupting vent(s), which can be either circular or linear. The former shape is often referred to as a central vent, whereas the latter is called a fissure.

Central-vent eruptions typically form roughly circular cratered hills or mountains that exhibit a cone-shape geometry when formed on land (Fig. 1.11b), but a more irregular heap-like geometry when formed in subaqueous or subglacial settings. On the other hand, fissures are typically delineated by a row of craters or cones (or both) or by jagged ridges when such eruptions occur under water or beneath a glacier (Fig. 1.11c).

Central volcanoes Central volcanoes are composite in the sense that they are constructed by repeated eruptions from a central vent system. They consist of multiple units of lava and tephra, ranging in composition from basalt to rhyolite. Central volcanoes in Iceland are grouped here into six categories on the basis of their average rise (i.e. slope), shape and arrangement of the main vents, and the composition of eruptives that dominate their successions (Table 1.4).

The first category includes the broad and gently sloping shield volcano,

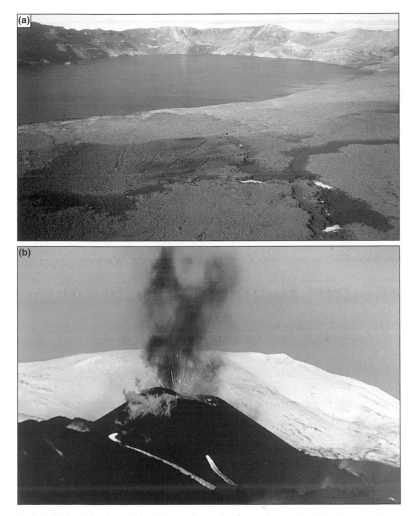

Figure 1.11 Examples of (a) caldera, (b) central vent and (c, overleaf) linear vent system.

capped by fairly large (7–15 km wide) summit **caldera** and characterized by eruption of basaltic magma (Fig. 1.12a). Most of the magma is erupted as fluid lava, which through recurring eruptions builds a shield-shape volcano. The Icelandic shield volcanoes bear a strong resemblance to those of other oceanic islands (e.g. Mauna Loa and Kilauea in Hawaii), although they are typically smaller. The second category includes the central volcanoes beneath the Vatnajökull icecap. These volcanoes have characteristics broadly similar to those of the shield volcanoes, although they exhibit a more erratic morphology because of the restrictions imposed by

27

the surrounding glacier. These volcanoes are also crowned by a caldera as opposed to a summit crater.

The third category, stratovolcanoes, represents the classic cone-shape volcanoes with steep outer slopes. These volcanoes are capped by a summit crater or a small caldera, often lined with lava domes along the rim slopes (Table 1.4, Fig. 1.12b). They feature a diverse style of volcanism, ranging from purely effusive mafic eruptions to explosive felsic eruptions, the latter producing widespread rhyolite fall deposits and pyroclastic flows. The fourth category consists of the Hekla volcano, grouped here as a special type of a stratovolcano because its vent system is distinctively linear and its products are predominantly andesite. Viewed from the southwest or northeast, the classic stratovolcano shape of Hekla is clear

Table 1.4 Classification scheme for composite central volcanoes in Iceland.

Environment	Type	Basal diameter	Slope	Vent shape	Magma types	Examples
Subaerial, subglacial	Shield volcano w/summit crater or caldera	24–38 km	≤8°	Circular	Basalt: minor andesite or rhyolite	Hofsjökull, Eyjafjallajökull, Mýrdalsjökull, Askja
Subglacial (highly irregular geometry)	Móberg volcano w/summit caldera	15–26 km	5–8° 20°	Circular	Basalt: minor andesite or rhyolite	Bárðarbunga, Grímsvötn, Kverkfjöll
Subaerial, subglacial	Stratovolcano w/summit crater or caldera	7–23 km	12–17°	Circular	Basalt–andesite–rhyolite	Snæfellsjökull, Snæfell, Öræfajökull
Subaerial	Stratovolcano ridge	6 km	15°	Linear	Andesite: less rhyolite and basalt	Hekla
Subaerial, subglacial	Caldera volcano	20 km	<3°	Multi-vent complex	Basalt: minor rhyolite and dacite	Krafla
Subaerial, subglacial	Caldera volcano	30 km	4°	Multi-vent complex	Rhyolite: minor basalt and andesite	Torfajökull

Table 1.5 Classification scheme for basalt volcanoes in Iceland.

	Eruption characteristics			Types of volcano	
Environment	Eruption type	Linear (long fissure)	Examples	Circular vent short fissure	Examples
Subaerial (magmatic)	Effusive (flood lava)	Spatter/scoria cone row (lava volume >10km³)	Laki, Eldgjá	Lava shield	Skjaldbreiður, Kollóttadyngja
	Effusive	Spatter cone row (lava volume <10km³)	Stampar, Prengslaborgir	Spatter cone (eldborg)	Eldborg at Mýrar, Búrfell (Heiðmörk)
	Effusive and explosive	Scoria cone row (lava volume <10km³)	Seyðishóla, Grábrók	Scoria cone (gjallgígur, gjallhóll)	Búðaklettur, Eldfell
Subaerial (hydromagmatic)	Explosive	Ash cone row or explosion crater row	Vatnaöldur, Veiðivötn	Tuff cone (hverfjall)	Hrossaborg, Hverfjall
	Explosive	Explosion chasm	Valagjá	Tuff ring and maar	Grænavatn, Víti
Subglacial and submarine	Effusive	Pillow lava ridge	Kverkfjallarani, Sveifluháls	Pillow lava cone	?
	Effusive–explosive	Móberg ridge, submarine ridge	Eldeyjarboði, Bláfjall	Móberg cone	Keilir, Jólnir
	Effusive–explosive – effusive	Table mountain	–	Seamount	Herðubreið
				Table mountain	Surtsey

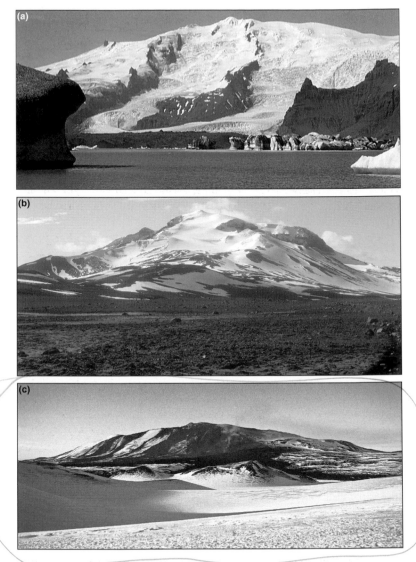

Figure 1.12 Central volcanoes: (a) shield volcano, (b) stratovolcano and (c) Hekla volcano.

(Fig. 1.12c). However, viewed from the northwest or southeast, it has an elongated form similar to that of a boat with its keel turned up. When it erupts, a 5 km-long fissure opens along its crest, creating a curtain of fire extending from the southwest shoulder, across the summit, to the north-west shoulder. Thus, repeated eruptions on the same fissure have built a ridge-shape stratovolcano.

VOLCANISM

In Iceland there are two active caldera volcanoes, which exemplify the last of the six categories of central volcanoes. These are Krafla, which mostly produces basalt, and Torfajökull, which mainly produces rhyolite (Table 1.4). These volcanoes are characterized by low sloping flanks and very large calderas (10–18 km wide) with multiple vents inside and along the rim of the caldera. They are dominated by effusive volcanism, although both have produced violent explosive eruptions.

Basalt volcanoes Fissure swarms typically erupt mafic magmas and are littered with basalt volcanoes, each representing one eruption (see Table 1.5, p. 29). Most commonly, the eruptions take place on a fissure and form volcanoes of the linear type. Volcanoes of the central-vent type are also present and are formed by eruptions that begin on short fissures but are rapidly reduced to one vent.

Most mafic eruptions produce both lava and tephra, although not always in the same proportion. Eruptions characterized by weakly splattering lava fountains form spatter cones. More explosive Strombolian eruptions feature more vigorously spraying fire fountains, producing scoria that accumulates around the vents to form scoria cones. These volcanic cones are typically surrounded by lava dispersed from the vent either through sealed lava tubes or open lava channels.

Lava shields are a special type of volcano and are thus named because they resemble a shield (skjöldur) lying face-up on the ground (Fig. 1.13a). They consist entirely of basaltic lava and are produced by prolonged effusive eruptions from a central vent. A central crater containing a lava lake spills lava in all directions and, with time, builds up a gently sloping (<5°) and remarkably symmetrical volcano, 1–20 km in diameter and 50–750 m high. The most recent example of a lava-shield eruption in Iceland is the effusive phase of the 1963–7 Surtsey eruption, which lasted for just over three years. The largest lava shields, such as Skjaldbreiður, are likely to have been formed by eruptions that lasted for several decades and possibly centuries.

Hydromagmatic eruptions often form deep explosion craters surrounded by low ash cones known as tuff rings or tuff cones (Fig. 1.13b). When the explosion crater is filled with water, it is referred to as a maar (Fig. 1.13c).

Eruptions under glaciers or in the sea are common in Iceland; the 1996 subglacial eruption at Gjálp in Vatnajökull and the 1963–7 submarine eruption of Surtsey are recent examples. Similar to other basaltic eruptions, subglacial and submarine eruptions can take place on a fissure or a circular vent.

When eruptions occur through vent(s) covered by a glacier several

31

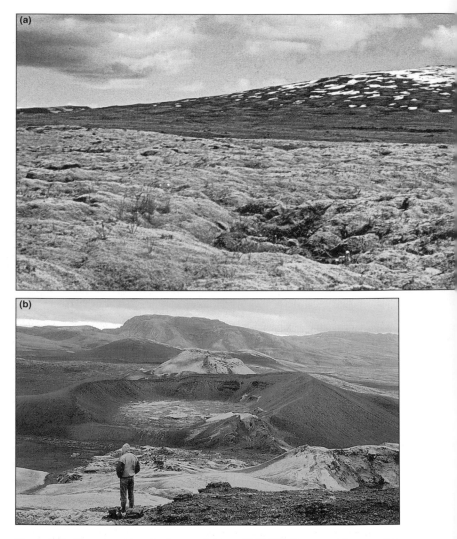

Figure 1.13   Examples of basalt volcanoes: **(a)** lava shield, **(b)** tuff cone, and **(c, opposite)** maar.

hundred metres thick, the magma melts the ice to form a water filled cavity at the glacier bottom and a matching depression in the glacial surface. Eventually the eruption may melt its way through the ice and form a lake. As long as there is sufficient pressure of water or ice (or both) to prevent explosive eruption, pillow lavas will accumulate around the vents (Fig. 1.14a). If the eruption stops at this stage, the resulting landform will be either a ridge or cone made up entirely of pillow lava.

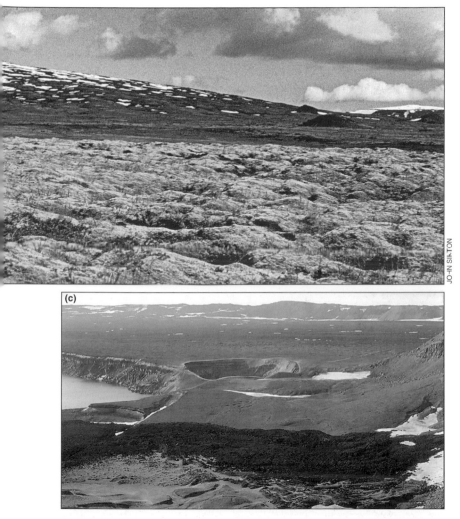

If the eruption continues, the volcano grows and the water depth will decrease. As a result, the water pressure decreases and the eruption enters an explosive hydromagmatic phase with formation of tephra that accumulates on top of the earlier-formed pillow lavas (Fig. 1.14b). The volcanic landform produced by eruptions that end at this stage is referred to as a móberg ridge or a móberg cone.

If the eruption continues, the volcanic edifice may grow large enough to prevent water from getting into the vents, and the eruption will become purely effusive, producing subaerial lava flows. If the tephra cone is still surrounded by water, the lava simply builds upon itself a lava delta made

Figure 1.14 Stages of growth of a subglacial eruption. **(a)** If the eruption stops at this stage, a pillow-lava ridge is formed; **(b)** a móberg ridge forms if it stops at this stage; **(c)** the formation of a table mountain.

of pillows and hyaloclastite, before continuing its advance (Fig. 1.14c). Volcanic landforms that have reached this stage form some of the most impressive volcanic structures in Iceland, known as table mountains (stapi). Submarine eruptions progress very much in the same manner as subglacial eruptions, and their volcanic architecture is identical (e.g. the Surtsey volcano, pp. 95–96).

Sedimentary rocks
Sedimentary rocks are a diverse family formed entirely by external processes operating at the Earth's surface. They are divided into three basic groups: chemical, organic and clastic.

Chemical sediments are formed by precipitation from water solutions because of supersaturation of a particular chemical agent, which can be induced by cooling of the solvent (e.g. formation of silica sinter from geothermal water) or evaporation of the water, as occurs when rock salt is formed. Silica sinter formed around geothermal springs is the only naturally occurring chemical sediment in Iceland.

The chief constituents of organic sediment are the remains of organisms: shells of freshwater and marine animals are the most common (e.g. limestones). Such deposits are not common in Iceland, but the silica shells of freshwater **diatoms** accumulate to form thick beds on lake floors (e.g. Lake Mývatn). Mass accumulation of plant debris results in the formation of sediments such as peat and lignite (surtarbrandur), the former being the main sediment fill of the many bogs in Iceland and the latter is a primitive form of coal, which occurs as distinctive seams in sedimentary beds between the lavas of the Tertiary Formation.

Clastic sediments are formed by the accumulation of debris consisting of minerals and rock fragments, which upon compaction and **lithification** become sedimentary rocks. Breakdown of existing rocks by weathering and erosion at the Earth's surface supplies the debris, which then is transported by gravity, wind, streams, glaciers or oceanic currents to be deposited elsewhere.

Similar to volcaniclastic rocks, clastic sediments are classified according to their textures. The principal classification is based on grain size – the distinguishing particle size of the sediment – and it includes familiar terms such as clay, silt, sand and breccia. When sediment is consolidated, the names change accordingly: for example, sand becomes sandstone.

Clast properties, such as angularity or roundness, are also used to distinguish between sediment types. For example, breccia consists of angular fragments that have not been modified during transport and thus represent immature sediments. Conversely, conglomerates consist of rounded pebbles, formed as angular fragments are chipped and ground down by grain collision during transport. Another important property of sediments is how well sorted the sediment is by grain size. Sediment is said to be well sorted when all of the fragments are roughly of the same size (e.g. beach sand) and poorly sorted when the sediment consists of fragments that vary greatly in size (e.g. till).

Clastic sediments are also classified according to means of transport, and take their name from the transport current or the environment of deposition: aeolian (wind blown), glacial, fluvial, lake or marine. In these depositional environments the sediment grain size and sorting reflect the carrying capacity of the transport current, which is determined by the flow velocity and the fluid viscosity. For example, although the wind can blow hard, air has low carrying capacity because of its extremely low fluid viscosity. Therefore, aeolian deposits are normally made up of particles of sand size or finer, which typically exhibit good sorting. Water has somewhat higher fluid viscosity and therefore is capable of carrying larger particles, especially when it is rushing through ravines or during flash floods. However, as we all know, rock sinks readily in water because of its higher density. Thus, in water-dominated systems such as fluvial, lake and marine environments, flow velocity and the sediment load principally determine the grain size and sorting of the sediment. The rule of thumb is that the grain size becomes larger with increasing flow velocity and the sorting becomes less as the sediment load increases. Consequently, boulders are the first to be deposited, followed by gravel, sand, silt and clay. This is well illustrated by many river systems. Where rivers flow fast in narrow and steep channels, the bottom sediments typically consist of boulders and stones. As the slope decreases and the river channel becomes wider, the river slows down and smaller grains, such as gravel (pebbles) and sand, settle to the bottom. Silt and clay-size grains settle out when the flow velocity is very low, hence their common occurrence in calm lake waters and deeper parts of the ocean.

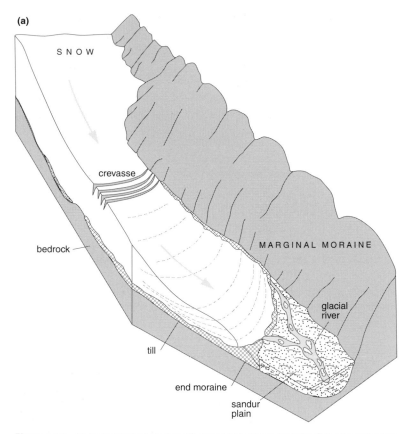

Figure 1.15 **(a)** Nomenclature of glacial features and glacial deposits. The transverse crevasses form above basement highs where the glacier is eroding the substrate (arrows indicate direction of flow). The till is formed as ground moraine at the glacier bottom, as marginal moraine along its sides and as terminal moraine at its front. Meltwater runs from the glaciers as rivers, forming a sandur plain in front of the glacier. **(b, opposite)** Example of an outlet or valley glacier, Skaftafellsjökull, South Iceland.

Glacial deposits differ significantly from those described above and are typically formed by mass deposition. Glaciers creep forwards as highly viscous fluid, similar to ductile tar or thick syrup, and thus are capable of carrying boulders of enormous sizes, as well as preventing the smaller particles from sinking. Therefore, glacial deposits are commonly an unstratified and chaotic mixture of clay- to boulder-size debris that is referred to collectively as till or tillite. Most glacial deposits are formed near the snout of the glacier, where its carrying capacity is greatly reduced. The basal layers of the glacier carry huge quantities of debris, which settles out to form sheet-like units of ground moraine at the glacier bottom and lateral moraines along its sides (Fig. 1.15a). At the glacier snout the debris piles up to form curved end moraines, which also mark the maximum advance of the glacier. Eskers are sinuous ridges of conglomerate that are deposited by rivers flowing within the glacier. When these rivers emerge from the glacier snout, they carry large quantities of debris, which is deposited onto the broad **sandur** plain in front of the glacier.

Sediments also form as the result of mass wasting on land or in the sea. Mass wasting involves downslope movement of particles or debris under the influence of gravity without the aid of a stream, glacier or wind. The resulting sediments typically consist of unstratified and chaotic deposits containing large angular stones and boulders floating in a matrix of sand- to clay-size material (Fig. 1.16a). The architecture of these deposit types is often very similar to that of till and lahar deposits and it is often very difficult to tell them apart. In such instances, these types of deposits are simply referred to as **diamictites**.

Talus slopes, an example of deposits generated by mass wasting, cover most mountain slopes in Iceland. They are formed over a long stretch of time by persistent accumulation of angular stones and blocks loosened by weathering from vertical rockfaces. Another example is rock (debris) ava- lanches, which are common along the steep glacially carved slopes of the fjord regions of Iceland. Rock avalanches occur when large chunks of steep

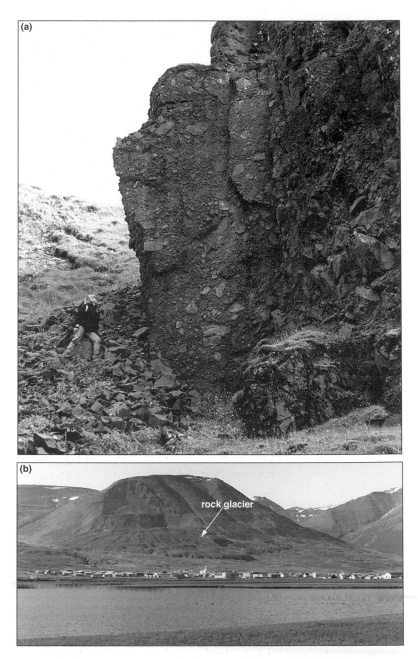

Figure 1.16 **(a)** Chaotic and unstratified diamictite in the Hreppar Formation in South Iceland, most probably formed by debris flow. **(b)** Rock glacier at Mt Hólshyrna above the town of Dalvík in North Iceland. Note the bowl-shape cirque above the rock glacier.

Figure 1.17 Striations marked in basalt lava by an eroding Plio-Pleistocene glacier. From the Hreppar Formation in South Iceland.

mountainsides or cliff faces cascade en masse into the valley below because of sudden slope failure. However, not all of the slump-like debris piles in these regions are avalanche deposits, but rather rock glaciers, formed when an icy rock mass creeps down slope with time (Fig. 1.16b).

The elements also contribute to denudation of the land by using the debris they carry as an erosional tool and thus are an important land-shaping factor. The dust carried by the wind polishes rock outcrops and exposed boulder surfaces. Rivers use the debris they carry to deepen and widen their channels by abrasion and in the process form V-shape valleys, gorges and canyons. Glaciers use the debris frozen into their base to grind down the land, as is evident in the striations they carve into the bedrock (Fig. 1.17). The glacial landscape is characterized by deep and steep-sided U-shape valleys surrounded by serrated ridges and hanging valleys.

Geothermal activity and hot springs
Geothermal energy is important for the economy in Iceland and is mainly used for domestic heating; just over 80 per cent of the population enjoy this facility. Schools have been deliberately located near geothermal areas to utilize this source of energy for heating and swimming pools.

Groundwater that percolates through cracks and voids in the top 1 2 km of the crust is heated in accordance with the temperature gradient of any particular area. In areas such as Iceland, where the temperature increases sharply with depth (50–200°C per 1000 m), the groundwater is

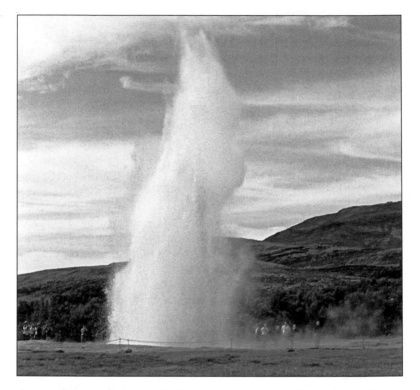

Figure 1.18   An erupting geyser, Strokkur, the most active gusher in the Geysir geothermal field.

quickly heated to elevated temperatures (70°C or more) and rises to the surface through fractures in the bedrock as geothermal water. The outlet vents of the geothermal water at the surface are known as hot springs (hver) and the area in which hot springs occur are known as geothermal areas. Geothermal areas are traditionally categorized on the basis of their overall average temperature: low-temperature areas, where the water is below the boiling temperature, and high-temperature areas, where the water is above it. Low-temperature geothermal areas typically occur in the regions bordering active volcanic zones, whereas the high-temperature areas are usually confined to active volcanoes. In addition, one can find cold or lukewarm carbonated springs, especially in the Snæfellsnes Volcanic Belt. These carbonated springs are sources of mineral water.

Some hot springs erupt by sending up a narrow column of water and steam (and occasionally rocks) high into the air (Fig. 1.18). Such erupting hot springs are known as geysers, named after the most famous of them, the Great Geysir in Haukadalur, South Iceland. This historic geyser has amused onlookers for centuries (since 1294) with its spectacular eruptions. In the past two to three decades the Great Geysir has been very quiet and it appeared to have retired from its previous activities. However, in June 2000, earthquakes reactivated it. Other Icelandic hot springs such as Grýla in Hveragerði and Strokkur in Haukadalur have also entertained the visitors with their frequent eruptions.

Tertiary, Quaternary, Holocene

The stratigraphical succession in Iceland records about 16 million years of geological history (Table 1.1), which appears trivial when compared to the 4000 million years of its nearest neighbour, Greenland. However, much can happen in 16 million years, especially in such a dynamic place as Iceland. Traditionally, the succession in Iceland is grouped into three major stratigraphical formations: the Tertiary Basalt (16–3.3 million years), the Plio-Pleistocene (3.3–0.7 million years) and the Upper Pleistocene Formation (<0.7 million years). As we shall see below, this stratigraphical division is based on the most distinctive volcanic rock types formed during each time interval, which broadly correlate with the changes in the climate through time.

The old rocks – the Tertiary Basalt Formation
The Tertiary Basalt Formation is the oldest stratigraphical formation in Iceland; it spans the time period from 16 to 3.3 million years ago (Table 1.1). It mainly appears in two large regions on either side of the active rift zones, covering about half of the country. In the east, the Tertiary Basalt Formation extends from Skaftafellsfjöll in the southeast across the eastern fjords to Bakkaflói in the northeast. In the west, it stretches from Hvalfjörður in the southwest, across Snæfellsnes and the western fjords, to Bárðardalur in North Iceland (see Fig. 1.2). The cumulative thickness of the Tertiary Basalt Formation is close to 10000 m. However, the true thickness of the succession at any location does not exceed 3000 m, because the vertical accumulation of volcanic and sedimentary rocks is coupled with outward spreading of the pile.

The original Tertiary landscape, which is not readily visible today, featured scattered central volcanoes, 300 m to more than 1000 m high, towering over broad and flat-lying lava plains. These plains were spotted with wetlands and dissected by the occasional gorge or river valley. Today we have an excellent cross-sectional view through the Tertiary succession, because of the extensive erosion by the Ice Age glaciers, a scene rarely available within the active volcanic zones. The most distinctive outcrop

41

feature of the Tertiary rocks is its layer-cake stratigraphy, where one basaltic lava flow is stacked onto another, forming gently dipping successions hundreds to thousands of metres thick. However, a closer inspection will show that there is more to it than this.

The Tertiary Basalt Formation consists mainly of volcanic rocks (> 85%) and was constructed by volcanotectonic processes similar to those operating in the currently active volcanic belts. Consequently, it features the same geological elements as the active volcanic zones. The exception is subglacial landforms, which are rare because the climate was considerably warmer. Thick basalt-lava series form 10–30 km-wide and 50–100 km-long lenticular bodies that represent volcanic systems. These lava piles are cut by many thin (1–10 m wide) dykes that strike parallel to the long axis of the systems. The density of these dykes is such that they often make up 3–15 per cent of the rock outcrop, and the dykes define distinctive swarms that transect the strata. In essence, the dyke swarms are the subsurface component of the fissure swarms in active volcanic systems (see Fig. 1.6). These dyke swarms are typically closely associated with clusters of andesite–rhyolite lava and tephra formations, marking the location of extinct central volcanoes. Altogether, 40 extinct volcanic systems have been identified in the Tertiary succession.

Although sedimentary rocks make up only about 10–15 per cent of the Tertiary succession, they are an important component because they best preserve information about the environment and the climate at the time. The most conspicuous are the red interbeds, so named because of their distinctive rusty red colour. Other types of sediments, such as finely bedded lacustrine siltstone and fluvial gravels, occur sporadically throughout the succession.

The red interbeds are ancient soils that were stained red by chemical weathering in the warm and humid climate of the Tertiary. Fossilized plant remains are often found in these ancient soils, which also contain fossil-rich lignite seams (i.e. primitive coal). About 50 genera of plant fossils have been found and the collection includes leaf imprints, fruit, seeds, pollen grains and tree moulds. The oldest flora, which includes trees such as swamp cypress, dawn redwood, vine and magnolia, indicates warm temperate climatic conditions in Iceland during the Miocene, similar to that of North Carolina or Portugal today. The younger Pliocene flora is characterized by conifers and broadleaf trees such as pine, spruce, alder and birch, indicating temperate climatic conditions and signifying the onset of the cooling that eventually led to the Quaternary Ice Age. The best known Tertiary fossil localities are in Northwest Iceland, including Húsavíkurkleif in Steingrímsfjörður and Surtarbrandsgil at Brjánslækur.

The middle-age rocks – the Plio-Pleistocene Formation
The Plio-Pleistocene Formation spans the time period of 3.3–0.7 million years ago (Table 1.1). It occupies a broad region on either side of the active rift zones and covers about a quarter of the surface of Iceland. On the east side it extends from Skaftártunga in the south to Langanes in the northeast. On the west side, it stretches from the Mt Esja region in the southwest to Skjálfandi in the north (see Fig. 1.2). The cumulative thickness of the formation is about 2000 m.

A large slice of the Plio-Pleistocene Formation, locally known as the Hreppar Formation, occupies the region between the active West and East Volcanic Zones in South Iceland. The formation also has two outliers, one at the Snæfellsnes Peninsula in West Iceland and another at Skagi in North Iceland (see Fig. 1.2). In both of the outliers, the Plio-Pleistocene volcanic products rest on glacially derived marine deposits, which in turn rest on much older rocks of the Tertiary Basalt Formation. The upper surface of the Tertiary succession is heavily eroded, indicating that a significant amount of time passed before the sediments covered it. Thus, here the boundary between the Tertiary and the Plio-Pleistocene formations is an unconformity, because there is a gap in the stratigraphical record across it.

As in the Tertiary, volcanic rocks of basaltic composition are the most widespread rock type in the Plio-Pleistocene succession. Subaerial basalt-lava flows make up a significant portion of the strata, but they alternate with subglacial formations such as pillow lavas, móberg breccias and tuffs. As before, andesite and rhyolite lavas and tephras are found in significant proportions in association with extinct central volcanoes.

Towards the end of the Tertiary (about 7 million years ago), the global climate had begun to deteriorate and the cooling trend that eventually led to the Quaternary Ice Age had truly set in. About 3.3 million years ago, when the first rocks of the Plio-Pleistocene Formation were forming, the "autumn" that preceded the frigid Ice Age "winter" had arrived in Iceland. Glaciers began to nucleate in the highlands, and the rivers grew larger because of increased precipitation. A progressive cooling resulted in incremental northward and westward growth of the Quaternary ice sheet from its nucleus in Southeast Iceland. By 2.5 million years ago it had covered more than half of the country and about 2.2 million years ago the whole of Iceland was covered by glaciers for the first time (Fig. 1.19). The Ice Age glaciation had arrived in full force.

However, this overall cooling trend was not uniform. It was characterized by frequent fluctuations between colder and warmer intervals. By 2.2 million years ago this fluctuation had developed into distinctive alternations between glacial stages, characterized by island-wide glaciation, and interglacial stages, when most of the land was free of ice. A total of nine

THE GEOLOGY OF ICELAND

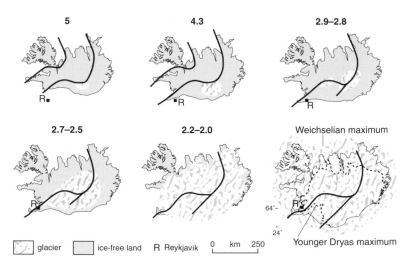

Figure 1.19   The growth of the Quaternary ice sheet in Iceland over the past five million years. The evolution of the Iceland rift system over the same period is also shown.

glacial and interglacial stages have been identified in the Plio-Pleistocene Formation, indicating that each glacial/interglacial cycle lasted on average for 180 000 years.

The changing climate had a marked effect on the landscape. Although the mechanisms of rifting and volcanism proceeded along the same general lines as in the Tertiary, the face of volcanic activity changed markedly. As the glaciers grew, subglacial eruptions became more frequent. Móberg ridges and table mountains were formed in ever-increasing volume. Likewise, while rivers and glaciers grew larger, their erosive might increased, forming deeper and broader valleys. Consequently, the Plio-Pleistocene landscape was typified by irregular topography, where jagged ridges and mountains towered over steep-sided valleys. The transformation of the climate during the Plio-Pleistocene is literally written in stone, because it resulted in the formation of new rock types not found in the Tertiary Basalt Formation.

During the interglacial stages, when the land was mostly ice free, the nature of the volcanic activity remained much as it did during the Tertiary, forming widespread lava flows and tephra deposits. On the other hand, during the cold spells, the glaciers put down various types of till and other glacial deposits, while subglacial eruptions produced vast amounts of pillow lavas, volcaniclastic breccias and tuffs. There is a noticeable increase in the proportion of clastic and volcaniclastic sediments up through the Plio-Pleistocene Formation. In the lower part, such deposits make up 15–30 per cent of the succession, increasing to 50–60 per cent in the upper part.

44

As the climate cooled, the red interbeds or soils become less conspicuous, while the expanding river systems were carrying forth more debris. The rivers dispersed the debris to the lowlands and to the sea, where it settled out to form fluvial, lacustrine or marine sediments in ever-increasing proportion.

The young rocks – the Upper Pleistocene Formation
The Upper Pleistocene Formation consists of rocks that are younger than 0.7 million years old, the distribution of which is for the most part confined to the active volcanic zones (see Fig. 1.2). The climate behaved in much the same way as during the latter half of the Plio-Pleistocene, with frequent alternations between colder glacial and warmer interglacial stages. Consequently, the Upper Pleistocene succession is typified by paired rock sequences, one formed under a thick glacier and the other on ice-free land.

The most prominent topographic features of the volcanic zones are from this period. This includes the central volcanoes, such as Katla and Askja, and the móberg ridges and table mountains that jut through the Holocene lava cover. The most youthful looking structures were formed during the Weichselian (most recent) glaciation, whereas more scrappy-looking structures that have suffered substantial glacial erosion date from previous glacial or interglacial stages. The younger table mountains constitute most impressive morphological structures (i.e. Herðubreið in North Iceland) and can be used to estimate the thickness of the Weichselian glacier (Fig. 1.20).

Five glacial periods have been identified in the past 0.7 million years, indicating 120 000–140 000 years' duration of each glacial/interglacial cycle. The Ice Age glacier reached its greatest extent at the maximum of the most recent glaciation, the Weichselian stage, which lasted from 120 000 to 10 000 years ago (Fig. 1.19). Today's glacial landscape is mainly the work of this last countrywide ice sheet.

At the maximum of the Weichselian glaciation, about 25 000–30 000 years ago, more than a kilometre of ice blanketed much of North America

Metres above sea level

Figure 1.20 Profile from Kverkfjöll volcano to the northern coast at Öxarfjörður, showing the elevation of subglacial móberg ridges and table mountains in the North Volcanic Zone and how it is used to reconstruct the thickness of the Weichselian glacier.

45

and Eurasia. The global sea level was 100–150 m lower than today, because large quantities of water were locked up as glacial ice. In Iceland, the ice sheet extended well beyond the present shores and covered the coastal shelf up to a distance of 130 km. About 18 000 years ago the bitter cold of the most recent glacial period was on the decline. The global climate became milder and the Weichselian glacier began to melt. This melting released large quantities of water, and sea level began to rise. As the glacier retreated, the sea followed in its path, inundating the land, which had been pushed down by the ice load.

However, this improvement in the climate was not without setbacks and we know of at least two periods when the cooling was great enough to see the glaciers recuperate and re-advance. These cold spells set in about 12 000 and 11 000 years ago and are known as the Older and Younger Dryas stages (see Table 1.1). The warm spell between them is the Allerød stage, a time when large parts of the lowlands in Iceland where at the bottom of the sea.

The start of the Holocene is usually set at the time when the global climate became as it is today, and in Iceland this occurred when the glacier of the Younger Dryas stage retreated from its end moraines about 9700 years ago (Fig. 1.19). The retreat of the Younger Dryas ice sheet was very rapid and most of it was removed by melting in only a few hundred years. The land now responded by rising extremely rapidly. Basically, it sprang up like a cork that has been pushed under water and then released. We know that in Iceland all of these events happened in a geological instant (i.e. less than 1200 years), because about 8500 years ago the Great Þjórsá lava flowed across ice-free and dry land, for a distance of 120 km, from its vents at Veiðivötn to the sea at Eyrabakki.

The Icelandic landscape of today was almost fully developed at the beginning of the Holocene. The exceptions are the river courses and canyons, which were formed after the melting of the glacier, and the active volcanic zones, which have been subjected to continuous modification by volcanic activity. Large parts of the sandur plains, which currently cover about 5 per cent of the land surface (5000 km²), were formed somewhat earlier by rivers emerging from the retreating glacier during the climatic transition. Since then, these plains have been modified and enlarged by the runoff from the present-day icecaps.

The pattern of volcanism during the Holocene has been similar to that of the Weichselian glaciation, although its habit changed significantly. All of the currently active volcanic systems have produced eruptions in the past 10 000 years. Most of the magma has been erupted as lava, which today covers about 12 000 km³ of the land surface. However, a significant amount (10%) of the basalt magma came up in explosive eruptions, as is evident in the many black and grey tephra layers in the soil cover. Many of these tephra layers are formed by subglacial eruptions, which have frequently been accompanied by catastrophic floods known as **jökulhlaups**. The brown and pale tephra layers in the soils are andesite and rhyolite fall deposits dispersed by explosive eruptions from the central volcanoes.

Volcanism in postglacial times has not been evenly distributed in space or time. The most productive area has been the East Volcanic Zone in Central-South Iceland, which has the highest eruption frequency and has produced the largest eruptions. Many of the large lava shields (e.g. Skjaldbreiður and Trölladyngja) and flood lavas (e.g. Þjórsá and Bárðardalur lavas) were formed during early Holocene times 8000–9500 years ago, showing that magma production was considerably greater than today.

The disappearing Weichselian glacier left behind barren and debris-covered land. The finer silt and sand fraction of the loose debris was picked up by the wind and dispersed, providing the raw material for soil formation. As the climate warmed, plants began to spread out from North and Northwest Iceland, where they had survived the Weichselian glaciation on ice-free areas. Birds and oceanic currents carrying plant seeds from the surrounding continents also introduced new species. Grasses, sedges and willow were the first to appear and were followed by birch, which spread rapidly and became the dominant species.

By the time the first Norseman arrived in Iceland in the ninth century, birch forest covered about 25 per cent of the country. The vegetation changed dramatically after the arrival of human beings. Not only did it result in the introduction of new species, mainly weeds, but also in the sudden expansion of grasses and sedges at the expense of birch and willow. A few centuries after the arrival of the first settlers, the woodlands had been reduced to a fraction of what they used to be. Today woodland and scrub cover 1 per cent of the country. This sudden and severe reduction in woodlands was in part attributable to the cultivation and domestication of the land, although it was also aggravated by gradual climatic cooling.

Chapter 2

The southwest

General overview

The geology of Southwest Iceland spans the past 3.2 million years, from
the very last stages of the Pliocene through the Quaternary to the present
day. The oldest rocks outcrop in and around Mt Esja in the north, the
succession becoming progressively younger to the south. The youngest
rocks outcrop along the axis of the Reykjanes Volcanic Belt (Fig. 2.1), which
connects the submarine Reykjanes Ridge and the West Volcanic Zone. The
belt consists of four volcanic systems, which are, from east to west, the
Hengill, Brennisteinsfjöll, Trölladyngja and Reykjanes systems. The systems have a northeasterly strike and extend across the peninsula.

The two excursions for this region provide a general overview of the
regional geology and brief descriptions of some exciting geological
features at selected stops. The first excursion is centred on the greater
Reykjavík area and the second one concerns the Reykjanes Peninsula.

Greater Reykjavík
A city on the margins of a mid-ocean rift

Situated on rather featureless rolling hills at the bottom of Faxaflói Bay,
greater Reykjavík may not appear to offer much excitement to geological
enthusiasts. However, there is more to it than catches the eye at first sight,
at least enough to kill a day or two, if one prefers rocks to shops. Situated
on the outer fringes of the Reykjanes Volcanic Belt (Fig. 2.1), it features
many examples of the contrasting forces that have shaped and moulded
the country in its recent geological past. These include glacial features, rift-
zone volcanics and tectonics. Some outcrops are within walking distance
from the city centre; others are within a short driving distance or can be
reached by city bus, plus a hike of a few kilometres.

Reykjavík ("smoky bay") was thus named by the first settler, Ingólfur
Arnarsson, in about 874 because of the columns of steam rising from

á house, æ **eye,** au feuille, ð **that,** ei day, é **yet,** hv quick, í seen, ö turn, ú soon, þ think

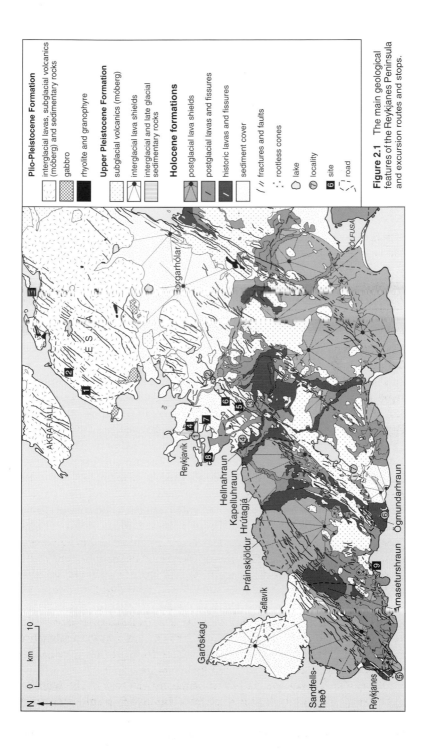

Figure 2.1 The main geological features of the Reykjanes Peninsula and excursion routes and stops.

Plio-Pleistocene Formation

- interglacial lavas, subglacial volcanics (móberg) and sedimentary rocks
- gabbro
- rhyolite and granophyre

Upper Pleistocene Formation

- subglacial volcanics (móberg)
- interglacial lava shields
- interglacial and late glacial sedimentary rocks

Holocene formations

- postglacial lava shields
- postglacial lavas and fissures
- historic lavas and fissures
- sediment cover
- fractures and faults
- rootless cones
- lake
- locality
- site
- road

geothermal springs in present-day Laugardalur. In the first decade of the twentieth century, these springs were used as washing pools, but they disappeared shortly after 1930 when their water was utilized for geothermal heating by the city of Reykjavík. The Laugardalur swimming pool, of course heated by geothermal water, is the only reminder of their past existence.

✳ Esja, the mountain that stands tall to the north of the city, is an essential component of the geological scenery of greater Reykjavík. Esja and smaller mountains around the town of Mosfellsbær are made up of the oldest rocks outcropping in the vicinity of the capital. The Esja succession covers about 1.3 million years, the oldest rocks (3.1 million years old) exposed at the foot of Esja in the west and the youngest (1.8 million years old) near the pale Móskarðshnjúkar that cap the mountain in the east. The strata dip gently to the southeast; their cumulative stratigraphical thickness is about 1650 m. Eastward and upward younging of the strata, and shallowing of its dip, reflect origin by spreading from the currently active West Volcanic Zone. The bulk of the Esja succession was formed by the now-extinct Kollafjörður and Stardalur volcanic systems (Fig. 2.1). However, the oldest rocks outcropping along the coast at Kjalarnes belong to yet another extinct volcano, the Hvalfjörður volcanic system (see Fig. 1.2b). The roots of their central volcanoes are exposed near the base of the mountain as large gabbroic intrusions. These gabbro bodies represent the tops of a solidified crustal magma chamber. In the past these chambers supplied magma to the eruptions at the three central volcanoes exposed in the Mt Esja stratigraphical succession. Many subvertical basaltic dykes cut the Esja succession, and some of those undoubtedly acted as feeder conduits to fissure eruptions. Such dykes are readily visible from the main highway along the northwestern slopes of Lokufjall and Eyrarfjall (Fig. 2.1: site 1).

✳ The upper part of the Esja succession consists of 11 relatively thick sequences of basaltic lavas, each paired with distinctive horizons of clastic sediments or subglacial volcanic deposits, or both. Initially, it was thought that each pair represented an interglacial and a glacial period. However, recent findings show that the lowest four sedimentary horizons were formed in a glacier-free environment. They are made up of fluvial conglomerates and sandstones, often intercalated with diamictite deposits laid down by debris flows or mudflows. Three of these units are exposed in accessible outcrops at Mt Eyrarfjall, the first in a cliff exposure at sea level, the second and the third at about 100 m and 200 m above sea level (Fig. 2.1: site 2).

The remaining sediment horizons are glacial in origin. The sediments consist of tillites and associated fluvial deposits, and they are often intercalated with subglacial móberg tuffs and pillow lavas. The oldest tillite in the Mt Esja succession, and therefore in Southwest Iceland, indicates full-scale glaciation at about 2.6 million years ago. The two oldest tillite beds are exposed in outcrops at Hvammsvík and at Fossá, close to Highway 1 on the south side of Hvalfjörður (Fig. 2.1: site 3).

The bedrock to most of greater Reykjavík is the so-called Grey Basalt (grágrýti in Icelandic), which is responsible for the rolling hills that characterize most of the urban landscape. The Grey Basalt is mainly covered by younger formations, but is exposed in road cuts and sporadic coastal and inland outcrops. It can also be seen in the walls of the Parliament House in downtown Reykjavík, because it is constructed from columnar blocks taken from one of the Grey Basalt lavas. At first the Grey Basalts of Reykjavík were thought to represent a single lava flowfield derived from the lava shield Borgarhólar at Mosfellsheiði to the northeast of the city. Now it is known that it consists of several lava flowfields formed by intermittent eruptions 100 000–200 000 years ago. Thin units of tillite, and fluvial and marine sediments, often separate the lava formations. The most widespread lava units are eroded remnants of lava shields, whereas the smaller lava units more produced by fissure eruptions.

A blanket of rather featureless till (ground moraine) laid down by the glaciers of the Weichselian period covers the bedrock. This is most apparent at higher elevations where the surface is usually littered with large boulders (urð). When the bedrock lava is exposed, its surface is usually polished and striated by glacial erosion. Sometimes such exposures exhibit the classical whaleback form of roches moutonnées. The striations and the whaleback structures are good indicators for the flow direction of the last Ice Age glaciers. Striated lava surfaces and ground moraines formed by the main Weichselian glacier are exposed at several other localities in and around Reykjavík, and a few of those are shown on Figure 2.1 (sites 4, 5, 6).

Below the peat bogs, at a depth of up to a few tens of metres, glacial deposits and glaciomarine sediments fill in the topographic depressions. Usually, at the bottom of this sediment sequence is basal till, or ground moraine, deposited by the advancing Weichselian ice sheet. The overlying end moraines, glaciofluvial and glaciomarine deposits were formed during the stage when the ice sheet was retreating from the area 9000–12 000 years ago. The land had subsided from the burden of the Weichselian glacier, so, when it withdrew, the ocean followed in its tract and submerged all of the lower-lying areas below the present-day 45 m altitude. This invasion of the ocean formed fossil-bearing deposits now exposed at several key localities (e.g. Fossvogur).

With the volcanic zone to the east and south of Reykjavík free of ice, the face of volcanism changed. Lava flows began to flow freely from lava shields and volcanic fissures, and several lavas made their way to the sea within the boundaries of greater Reykjavík (see Fig. 2.1).

Locality 2.1 Deglaciation history of Reykjavík – the Fossvogur sediments and other glacial deposits

The Fossvogur sediments are one of the classic geological sites in Iceland that have attracted geologists and other naturalists for more than 150 years, ever since they were first investigated in 1836 by the French naturalist Louis Eugéne Robert. These glacial and fossiliferous glaciomarine deposits tell an intriguing story, which through time has fostered widely varying ideas about their origin and age. Until recently, it was generally believed that the sediments were formed during the most recent interglacial (i.e. Eemian stage) 70 000–130 000 years ago. Now we know that the deposits are much younger and were formed during the demise of the Weichselian glacier. Radiometric ^{14}C dating shows that they were laid down 10 500–11 500 years ago, during the Allerød stage. Thus, the deposits are an integral part of the deglaciation history in greater Reykjavík.

In short, the deposits of the Fossvogur sequence illustrate a cyclical climatic change that began with deposition of the tillite during the Weichselian glaciation. This was followed by deposition of fossiliferous marine sediments as a result of partial deglaciation and sea invasion during the warming of the Allerød stage. Shortlived cooling in mid-Allerød time temporarily interrupted this trend and resulted in a brief advance of the Fossvogur tidewater glacier and formation of submarine debris flows (Fig. 2.2a). The sequence is capped by till and other glacial deposits formed by re-advancing glaciers during the cold snap of the Younger Dryas stage (10 500–9800 years ago; Fig. 2.2b).

The Fossvogur sediment sequence is exposed in a series of coastal outcrops, 2–5 m above the land surface, that extend for about 2 km along the northern shoreline of the inlet from which the sequence takes its name. Nautólfsvík, directly south of Reykjavík airport, provides the best beach access to the Fossvogur outcrops. Those who wish to examine the outcrops from the beach are advised to do so at low tide (tidal range is 1.7–3.8 m) and put on waterproof footwear. Alternatively, one can follow the path at the top of the coastal cliffs to examine the outcrops. In outcrop the arrangement of the diamictite and fossiliferous mudstones to sandstones that make up the Fossvogur sediment sequence is rather complex (Fig. 2.3a) and its history is best described by a composite graphic log where the sediment units, labelled 1 to 11, are arranged in stratigraphical order (Fig. 2.3b).

The bedrock to the Fossvogur section is the Grey Basalt lava (unit 1) and it pokes out as whaleback rises with glacially striated surfaces. Massive

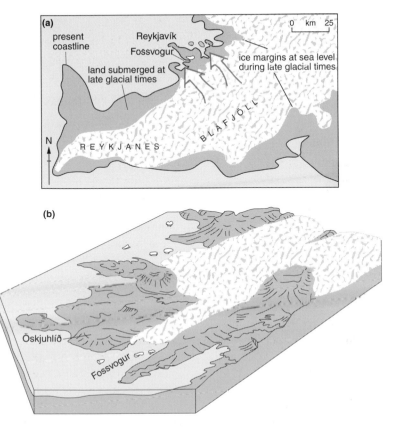

Figure 2.2 Glaciers in Southwest Iceland during the Allerød stage: **(a)** extent of the Reykjanes glacier during the Allerød stage; **(b)** snapshots of the Reykjavík area during the times of the Fossvogur tidal glacier.

diamictite (unit 2) containing striated boulders often fills the depression between the lava rises. Together, these units represent bedrock erosion and deposition of basal till by the main advance of the Weichselian glacier.

The overlying deposits (unit 3) consist of beach conglomerates. These deposits mark the transition from a glacial to a marine environment, as the Weichselian glacier retreated and the sea began its invasion. The super-imposed fossiliferous mudstone and siltstone deposits (units 4 and 5) record continued marine invasion in the early Allerød time. The lower laminated mudstone units were formed as tidal flats. The upper massive-to-bedded sandy siltstone units and their fossilized fauna were formed in somewhat deeper water in a protected fjord environment. The abundance of **dropstones** in these units indicates that the glacier was not far away. These dropstones were carried by drifting icebergs originating from a

53

Figure 2.3 **(a)** Coastal section showing part of the Fossvogur sediment sequence in and around Nautólfsvík; **(b)** composite graphic log of the Fossvogur sediment sequence showing the stratigraphical order of the main sediment units and their interpreted origin.

calving glacier at the head of the fjord and dropped to the bottom of the sea as the ice melted.

Upwards, unit 5 grades into fossiliferous silty sandstone (unit 6), indicating continued deposition in a submarine environment. This sequence is interrupted by beds of crudely stratified coarse-grain diamictites (units 7 and 8), which were formed by intermittent subaqueous debris and mudflows. These units contain glacially striated basalt boulders and mudstone clasts picked up by the flows through erosion of the underlying strata. This sequence is capped by a strong erosional surface and glacial deposits (unit 9) formed by temporary advance of the tidewater glacier into the old Fossvogur fjord. The morainal hills at the head of the current Fossvogur inlet were also formed at this time.

The fossiliferous siltstones and sandstones of unit 10 indicate renewed retreat of the glacier and sea invasion towards the end of the Allerød stage. The bedded gravels and boulder beds (unit 11) at the top of the sequence formed when the area was once again covered with a glacier during the cold snap of the Younger Dryas stage. A raised boulder-beach deposit is found in Öskjuhlíð, the hill north of Fossvogur, at 43 m above sea level. It represents the highest stand of the sea in the wake of the Younger Dryas glacier.

To complete the history of deglaciation in the greater Reykjavík area, we must explore other sites. Fossiliferous glaciomarine sediments of similar character and identical age to the Fossvogur beds are exposed in an outcrop along the north shore of Kópavogur inlet (see Fig. 2.1: site 7). Near the bottom of the inlet these marine deposits are overlain by glaciofluvial sand and gravels, and an abutting end moraine that can be followed to the south up the slopes of the Arnarnes point. Farther out, a well defined end moraine extends across Álftanes (see Fig. 2.1: site 8), which clearly represents a temporary standstill of a major glacier. This moraine was formed after the warm spell of the Allerød stage and it represents the maximum glacial stand during the Younger Dryas stage (about 10 200 years ago). When the Holocene warming set in for good, the glacier retreated rapidly away from these moraines, thereby leaving them as a record of its previous activities.

Locality 2.2 The Rauðhólar rootless-cone group at Elliðavatn
Follow Miklubraut, the main east–west artery through Reykjavík, east towards the suburb Árbær, then turn south onto Suðurlandsvegur, the branch of Highway 1 that leads to the town of Selfoss, and follow it for about 4 km. There, on the eastern outskirts of Reykjavík just south of Highway 1 is a cluster of rather inconspicuous reddish hills that can be easily overlooked (see Fig. 2.1). These are the Rauðhólar ("red mounds"), which are an excellent example of rootless-cone groups, a volcanic landform that has been identified only in Iceland and on the planet Mars.

Rootless-cone groups consist of closely packed cones that rest directly on the associated lava flow. Cone dimensions vary from 2 m to 40 m high and 5 m to 450 m wide. They occur in tube-fed pahoehoe flows where the lava advanced over wetlands, such as shallow lakes or swamps. The area of individual cone groups is most commonly between 1 and 10 km^2, but the largest cone group, Landbrotshólar, covers 150 km^2. Rootless cones are formed by hydromagmatic eruptions caused by explosive interaction between molten lava and water-saturated substrate. Their cones represent volcanic vents that have lateral feeders, which are the lava tubes of pahoehoe flows. Thus, they differ from normal vents in that their feeders are not rooted deep within the crust, hence the name "rootless cones".

The naturalists Sveinn Pálsson (1793) and Eugéne Robert (1840) were the first to suggest a secondary origin for the rootless-cone groups in Iceland. Another century passed before this idea was firmly established, when the Icelandic volcanologist Sigurður Þórarinsson showed that the Mývatn cone group in North Iceland was indeed formed by rootless eruptions. However, the exact mechanism of their formation was unravelled only recently by studies of the Rauðhólar cone group and we therefore outline the results in some detail.

Rauðhólar occur as a well confined field ($1.2\,km^2$) of small scoria and spatter cones within a 2 km-wide lava branch derived from the 4700-year-old Leitin lava shield of the Brennisteinsfjöll volcanic system. This branch of the lava extends some 27 km from its source at Bláfjöll to the sea at Elliðavogur in Reykjavík (see Fig. 2.1). Several smaller cone groups occur in the Leitin lava, including the minute but well known Tröllabörn cone group about 5 km east of Rauðhólar. The Rauðhólar cone group is located where the lava flowed into and covered a small lake just north of present-day Lake Elliðavatn. The cone group is confined to a raised (5 m high) scoria platform that rests on 7 m of solid lava. In turn, the lava rests directly on 1–2 m-thick mudstone that covered the floor of the lake before the lava arrived.

The Rauðhólar cone group originally featured about 150 cones and craters, but roughly one third of the cone group was excavated for road ballast in the middle of the twentieth century. The largest cones are located within the northern half, which is characterized by a cluster of closely packed near-circular cones (Fig. 2.4a). The random arrangement of cones and craters is obvious when one attempts to isolate individual structures. The German volcanologist Maurice van Komorowitz described the group in 1912 and his map is the only available source that shows the original size and distribution of the Rauðhólar cones. The largest cone of the group had a basal diameter of 212 m and rose 22 m above the surrounding lava surface. The smallest cones, still seen in the southern part of the cone group, are only a few metres high and exhibit **hornito**-like forms.

The excavated part provides excellent exposure to examine the internal structure of the cone group and individual cones. These structures, which provide important clues about the formation of the cone group, are shown schematically on the reconstructed profile on Figure 2.4b. Each cone has a crudely funnel-shape conduit (not visible) extending from the base of the flow and up through the coherent lava, terminating in a bowl-shape crater. Another noteworthy feature is that later cones lap onto and sometimes partly conceal earlier cones. Most importantly, individual cones exhibit distinctive internal layering, showing that each cone was built from multiple explosions during a period of sustained activity.

Figure 2.4 **(a)** Rauðhólar rootless-cone group, showing the original arrangement of pristine cones in the western sector of the group. **(b)** Reconstructed cross section through the Rauðhólar cone group, showing the general arrangements of cones and their characteristic internal structures.

In section, the cone ramparts typically feature a well bedded lower sequence of ash and scoria and a crudely bedded upper sequence of coarser-grain agglutinates. The lower sequence consists of 0.2–0.6 m-thick layers composed of highly fragmented lapilli scoria alternating with thinner (<0.2 m) crudely laminated beds of red-baked lacustrine silt and black ash. The scoria layers also contain, in variable abundance, lumps of red-baked mud and clasts that have cores of either vesicular lava fragments or cooked mud armoured with a skin of quenched lava. The upper sequence features multiple 0.5–1.5 m-thick layers of brick-red agglutinates, mainly composed of spatter bombs with fluidal surfaces and variable twisted or distorted shapes. The distinctive brick-red colour of the deposit comes from a micrometre-thick coating of red-baked mud that was welded to the surface of the clasts while hot. Lapilli scoria is present in only minor amounts and the abundance of mud lumps is significantly

57

lower. A 1–2 m-thick lava-like layer of welded spatter often caps the agglu-tinate beds. The upward increase in grain size within the deposits emu-lates the attenuation of explosive power during an individual rootless eruption and, importantly, this change goes hand in hand with a decrease in the abundance of lacustrine mud incorporated into the deposits by the explosive activity.

When considering the eruption mechanism that produced Rauðhólar and other rootless-cone groups, the following facts need to be kept in mind. First, although the cone groups rest directly on top of the host lava, their cones show no evidence of having been deformed or modified by moving lava. Secondly, the occurrence of lacustrine muds as distinct beds within the lower sequence is a clear indication of their involvement in gen-erating steam explosions by providing the water (i.e. the coolant). Obvi-ously the lava was the fuel. Thirdly, the internal stratification of individual cones shows that they were formed by multiple explosions of decreasing intensity, whereas overlapping arrangements of cones within the group imply a certain time sequence and duration for the explosive activity. Fourthly, we have to consider the apparently random distribution of the cones and the fact that the lava continued its advance despite the rootless eruptions. In other words, we have to get the lava across the lake bed and at the same time initiate rootless hydromagmatic eruptions through con-tact between hot lava and water, forming cones on top of the lava that are in no way modified by its movements.

These conditions can be met if the lava is a tube-fed pahoehoe flow (Fig. 2.5). Initially, the lava enters the lake as relatively small pahoehoe lobes from a set of tube entries at the active lava fronts. The insulating crust seals the lobe interiors from the water, which inflate and expand laterally in response to continued injection of lava. The internal lava tubes are thus extended to new lake entries and the process repeats itself, extending the lava farther into the lake. Another consequence of this process is that, as the lava behind the active flow fronts increases in thickness by inflation, it begins to sink into the soft mud on the lake floor. However, the subsid-ence is not uniform and, when cracks open in the base of the lava below the internal pathways, the glowing hot lava flows straight into the water-saturated mud and initiates the steam explosions. If the explosions are powerful enough, they burst through the overlying lava to emerge as root-less eruptions that build cones around the vents (Fig. 2.5a). At any site, the eruption stops when the supply of water (i n mud) or lava runs out. As the flow migrates across the lake bed, the explosive vents follow and, in doing so, gradually build a group of rootless cones on top of the lava. When the lava eventually reaches the opposite shore, it continues its advance in the same manner as it did before its encounter with the lake bed.

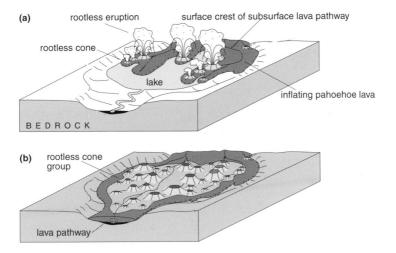

Figure 2.5 Formation of the rootless-cone groups. **(a)** The lava has covered the basin and has thickened because of lava inflation. At this stage the lava is transported across the basin in narrow lava tubes under insulating crust. Cracks in the tube floors form entries for the hot lava into the underlying water-logged lake sediments, initiating series of rootless eruptions. **(b)** A fully developed rootless-cone group.

Locality 2.3 The Hjallar fault system and the Búrfell spatter-cone volcano
Travelling south through Rauðhólar, follow the unsealed road that wiggles around the eastern side of Elliðavatn, and then the signs to Heiðmörk.

The basement rocks of Heiðmörk are the Grey Basalt lavas. The lavas are partly covered by ground moraine of the Weichselian glacier, but when exposed they exhibit examples of glacially striated surfaces. On the eastern side of Heiðmörk, the Grey Basalt is covered by a few younger Holocene lava flows. The one that comes closest to the road is the Strípshraun flow, a basalt lava with abundant large olivine phenocrysts (> 10 per cent by volume). This lava flow is 9 km long and it originated at a cone near Þríhnjúkar that has the peculiar name Eyra ("ear").

We are now within the Hjallar fault system, which consists of northeast-trending fissures and faults defining a series of narrow **graben** and **horst**. It marks the northern end of the fissure swarm associated with the Krýsuvík volcanic system (see Fig. 1.5b). Hjallar is actually the name of a 5 km-long scarp marking the northwestern margins of the fault system, which is characterized by stepwise normal faulting to the east, with a cumulative throw of 65 m. The fault system continues eastwards until it disappears under younger Holocene lava flows. Some of the fissures and faults cut across the Holocene lavas, but reveal much smaller fault displacement, in the order of 1–12 m. This shows that the faults are not formed in a single event, but grow by repeated movement on the same fracture over a substantial period of time. The fissures and the faults are absent in the historical lavas nearby, which suggests a lull in the tectonic activity in this part of the fissure swarm over the past 1100 years. Although the system may lay dormant for now, it is still considered active.

Búrfell is a spatter-cone volcano situated near the middle of the Hjallar fault system about 1.8 km southeast of the point where the Heiðmörk road turns sharply to the northwest. The cone is well within walking distance from the road, an appealing and informative hike because the trail follows a drained lava channel that was full to the brim of red-glowing lava about 7200 years ago.

Búrfell is a near-circular cone with ramparts made up almost entirely of spatter ejected by lava fountains. The crater rims rise up to 80 m above their surroundings and circumscribe a 140 m-wide and 60 m-deep crater, which contained a lava lake during the eruption. One of the northeast-trending Hjallar faults dissects the cone, and the eastern cone rims have dropped 2–5 m because of movements on this fault. The fault continues into the basement rocks to the north and south of the cone, where the vertical displacement across the fault is much greater (about 12 m). Thus, the Búrfell magma followed an existing fault to the surface and that additional displacement occurred on this fault some time after the cone was built.

The lava that emanated from the Búrfell cone flowed to the northwest for about 12 km before entering the ocean in Hafnarfjörður and Skerja-fjörður. The geological name of the flow is the Búrfell Lava, but, in line with established Icelandic tradition, various sectors of the flow take names from nearby landmarks. Thus, locals may refer to the Búrfell Lava by up to nine different names, much to the annoyance of visitors (Fig. 2.6). One sector, Gálgahraun ("gallows lava"), reminds us of a rather unpleasant part of Iceland's history. As the name implies, it hosted the gibbets used to execute serious offenders in the past.

The Búrfell Lava consists of three main branches, and their relative positions tell us how the eruption progressed. The first to form was the Kaldár-sel branch. At the beginning of the eruption, the lava flowed in an open channel to the south, feeding a parasitic (secondary) lava pond that formed at present-day Kringlóttagjá (Fig. 2.6). By transport via internal lava pathways or tubes, the Kringlóttagjá pond fed inflating pahoehoe lavas advancing down towards Kaldársel and then to the northwest through the valley between Sléttuhlíð and Stórhöfði, where they disappear under the younger Gráhelluhraun branch of the Búrfell Lava. The Kaldársel branch disappears under younger lavas to the south. The mechanism of lava inflation was first recognized by Sir George Stuart Mackenzie, Sir Henry Holland and Dr Richard Bright, when they examined the Kaldársel branch in 1810, a discovery that went unnoticed for the subsequent 180 years.

Figure 2.6 Búrfell spatter cone and lava flow, showing the key structures, distribution and local names for the lava. The flow covers a total area of 18 km² and its volume is 0.35 km³.

As the eruption continued, the Búrfell cone grew in height and so did the level of the lava lake in the crater. As a result, the lava breached the cone rims in the southwest, cascading down the outer slope and spreading lava westwards from the cone and forming Smyrlabúðarhraun. There it split into two parallel branches. The southern one was formed as the lava flowed through the corridor between Sléttuhlíð and Setbergshlíð, forming Gráhelluhraun, and the narrow strip of lava that extends all the way to downtown Hafnarfjörður, where it disappears under the lavas from the northern branch. The northern one (the largest branch) followed a direct northwest passage between Vífilstaðarhlíð and Setbergshlíð, then spread westwards to the coast in Hafnarfjörður and Skerjafjörður. In the process the lava blocked a small valley and formed the small Lake Urriðakotsvatn.

61

The 3.5 km-long Búrfellsgjá lava channel was formed during this stage of the eruption, feeding lava to the northern branch. The surface of these two lava branches is to large extent covered by rubble. In some places these branches exhibit structures, such as lava tubes and classic pahoehoe flow margins, indicative of lava inflation (e.g. Fig. 2.6: sites 1 and 2).

Locality 2.4 Tvíbollahraun and Kapelluhraun, the only historical lava flows in the greater Reykjavik area
On the outskirts of Hafnarfjörður, driving west on Highway 41, are the only historical lava flows within the urban limits of greater Reykjavík (see Fig. 2.1). The one closer to Hafnarfjörður is called Hellnahraun ("pavement lava") and was formed by an eruption at the Tvíbollar cone row in about 950. The vents are out of sight, at a locality called Grindarskörð, and are within the Brennisteinsfjöll volcanic system. Thus, the lava travelled 17 km before coming to halt here at Hvaleyrarholt, flowing across the Krýsuvík volcanic system in the process (see Fig. 1.5). It is pahoehoe lava and it features exceptionally well preserved surface structures, such as tumuli and lava-rise pits, indicative of lava inflation. The tumuli are easily recognizable, as mounds bounded by raised crustal slabs and dissected by a central cleft. The other one is Kapelluhraun, also known as Nýjahraun ("new lava"), and is derived from a cone row located at the foothill of Undirhlíðar, about 7.5 km directly east of the main road. Kapelluhraun is an aa lava flow and was formed by an eruption in the twelfth century during a major volcanotectonic episode on the Trölladyngja volcanic system. The Ögmundarhraun lava near Grindavík was produced by the same episode (see also p. 69).

Reykjanes–Grindavík–Kleifarvatn
The architecture of a mid-ocean ridge

To begin this excursion from Reykjavík, simply follow the route to the international airport at Keflavík, head south through town and then west on Highway 41.

Despite its inhospitable appearance, the Reykjanes Peninsula is seventh heaven for geologists, especially tectonics and volcano enthusiasts. Here vegetation cover definitely does not impede exposure, and rocks outcrop at every footstep. It is incomparable in the sense that nowhere on Earth is the intricate architecture of a mid-ocean ridge system better exposed or as accessible for intimate inspection. Its inimitability does not end there, because it is probably the best example available to us that mimics the structure of the very first landmasses formed on Earth and thus

can be viewed as a window into the deepest past of geological history.
The Reykjanes Volcanic Belt consists of four northeast-trending volcanic systems arranged in a step-wise fashion across the peninsula. The
mountain range along the centre of the peninsula roughly delineates the
axis of the belt. The steady westward drop in altitude from the Hengill volcanic system, at 803 m above sea level, to the Reykjanes system (243 m),
represents decreasing influence of the mantle plume. The buoyancy
effects of the plume are gradually tapering off with increasing distance
from the centre of the mantle plume. The mountain range consists of subglacial and submarine volcanic structures, such as móberg ridges and
table mountains, flanked, and in parts overlain, by subaerial lavas formed
during the Holocene and to a lesser extent in historical times (see Fig. 2.1).
Disregarding the area from Vogarstapi to Garðskagi, the northwestern
part of Reykjanes Peninsula consists of three abutting Holocene lava
shields, which are, in order from east to west, the Hrútagjá, Þráinskjöldur
and Sandfellshæð. A similar arrangement is found on the southeast side
of the peninsula, where six lava shields occupy similar positions (Geit-
ahlíð, Herdísarvík, Selvogsheiði, Heiðin há, Leitin and Tröllahlíð).
Hrútagjá and Þráinskjöldur are abutting half shields with their sum-
mits at 200–240 m above sea level and tugged onto northwest slopes of the
móberg mountain range. They produced pahoehoe lavas that spread radi-
ally northwards and form the 20 km-long coastline between Straumsvík
and Vogarstapi, some 10–11 km from the source vents. The Hrútagjá
shield lavas formed some of the largest tumuli found in Icelandic lavas,
whereas Þráinskjöldur is heavily dissected by fissures and faults formed
by rifting on the Reykjanes volcanic system. These faults are easily seen
from the road in the form of northeast-trending linear scarps on the lower
slopes of the shield. Farther up slope is the pyramid-shape móberg cone
Keilir, an island amid an ocean of lava.
The third lava shield, Sandfellshæð, has a more regular shape and
covers a large portion of the southwest corner of the peninsula. It is an
exceptionally gently sloping shield (<3°) that rises to 90 m above sea level
at the summit and its lavas cover 120 km^2. Sandfellshæð has a well formed
summit crater, 450 m wide and 20 m deep, which is located about 4–5 km
east of the road (see Fig. 2.1).
All three shields were formed at the beginning of the Holocene about
9000–10 000 years ago and their cumulative volume is 15 km^3, accounting
for more than 75 per cent of the volume of magma erupted onto the surface
in this sector of the peninsula in the past 10 000 years. They are also the
principal agents responsible for raising the area out of the sea. The original
distribution of the shield lavas is partly obscured by younger Holocene
and historical lava flows; examples include the historic Hellnahraun and

Kapelluhraun lavas (mentioned earlier), as well as the youthful-looking Afstapahraun lava flow, which entered the sea some 7.5 km farther to the west. This 14 km-long aa lava originated at a 3 km-long cone row near Tröl-ladyngja and it fills in the low where the flows of the Hrútagjá and Þráin-skjöldur lava shields meet.

The Quaternary basement that outcrops from Vogarstapi across Mið-nes to Garðskagi is the oldest rock formation west of Reykjavík. It is mostly composed of glacially eroded lavas formed by one lava shield volcano or more. The easternmost part (i.e. Vogarstapi) is heavily dissected by young fissures and faults formed in association with activity on the Reykjanes volcanic system. Directly south of Vogarstapi are the remains of Stapafell and Þórðarfell, two submarine volcanoes composed of strongly olivine phyric pillow lava and associated volcaniclastic deposits formed at the end of the Weichselian glaciation. These volcanoes have been removed almost completely by excavation. Rauðimelur is the name of a 3 km-long spit bar that extends to the northwest from Stapafell. The bar was formed by marine erosion of two volcanoes, as is evident from its composition – basalt clasts mixed in with olivine crystals.

Locality 2.5 Mid-ocean ridge rising out of the sea
If you travel west on Highway 41 towards the international airport at Keflavík, as you pass the town of Njarðvík follow the signs to the town of Hafnir (Road 44). Head directly south on Road 425 through Hafnir and to the salt factory at Reykjanes. From there follow the signs to Reykjanesviti (the lighthouse at Reykjanes) and, passing the lighthouse, you have reached the destination (see Fig. 2.1).

Reykjanes is the southwestern outpost of Iceland and the point where the Mid-Atlantic Ridge rises out of the sea. On land the ridge crest (i.e. the rift valley) is defined as a distinct fault zone delineating a shallow 10 km-wide composite graben structure trending northeast from Reykjanes to Vatnsleysuströnd. Along the centre of the rift is a series of 40–100 m-high steep móberg hills composed of pillow lava and hydroclastic tephra and breccia, including Bæjarfell, where the present lighthouse stands, and Val-ahnúkur, where the foundations of the old lighthouse can still be seen. On both sides, lava shields border the móberg hills, which in turn are partly covered by younger Holocene and historical lava flows (see Fig. 2.1). But not all of Reykjanes is volcanic. The cove south of Valahnúkur features a spectacular boulder beach formed by the rampant storms of the North Atlantic, and farther to the north the black sand dunes at Stóra-Sandvík are a blunt reminder that the wind blows fearlessly through Reykjanes. It also hosts a small (~1 km²) high-temperature geothermal area with small steaming vents and bubbling solfataras. However, these hot springs are only a shadow of what they used to be, partly because the area is now utilized for energy by the nearby salt factory.

The móberg mountains of Sýrfell, Bæjarfell and Valahnúkur were formed by submarine fissure eruptions when sea level was standing about 70 m higher than today during the waning stages of the Weichselian glaciation. These móberg mountains consist of pillow lava, breccias and tuffs, which in the case of Sýrfell are capped by small scoria cones and thin lava flows, indicating that the volcano rose out of the sea to form a small island. However, Valahnúkur volcano seems to have remained fully submerged during its period of activity, as it features a basal sequence of pillows capped by a breccia unit and another pillow-lava sequence. The hill itself is split in the middle by a 20 m-wide graben structure, and the walls on either side provide an excellent outcrop of the pillows and the breccia units (Fig. 2.7a). The youngest tephra layers in the **regolith** that laps on to the eastern slopes of Valahnúkur are from a volcanotectonic episode that raged at Reykjanes in the thirteenth century, suggesting that this graben structure is very young and was formed in historical times.

The youngest volcanic formation at Reykjanes is the 4.5 km-long Yngri Stampar cone row, along with the lavas and the tuff it produced, which now cover the northern part of the point (see Fig. 2.1). These were formed in the so-called Reykjanes Fires, a major volcanotectonic episode within the Reykjanes system between 1210 and 1240.

Vatnsfell is a rather inconspicuous rise on the coast north of Valahnúkur, but is worth a visit for those interested in having a closer look at the internal make-up of tuff cones. It is made up of lapilli tuff deposits, which are the remnants of two tuff cones formed by submarine explosive eruptions in the early stages of the Reykjanes Fires in about 1210. The internal structures of these cones are best exposed in outcrops opposite Karl, the sea stack rising about 300 m off shore (Fig. 2.7c). The older cone, which makes up the bulk of Vatnsfell proper, was 30 m high and 650 m wide, with the crater located 100 m off shore. The lower part of the older tuff cone consists of wavy to cross-bedded ash deposits formed by repeated base surges, whereas the upper part consists of coarser-grain lapilli tuff composed of alternating base surge and tephra fall units. On either side of Vatnsfell, the older cone is capped and flanked by the deposits from the younger cone, which consists of fine-grain lapilli tuff with repeated pairs of base-surge and tephra-fall layers (Fig. 2.7b). Many impact craters formed by ballistic blocks ejected from the craters during the eruptions commonly disrupt bedding in both cones. The younger tuff cone was a much larger structure than its earlier counterpart, at least 55 m high and with a basal diameter of about 1600 m. The main crater was located about 400 m off shore near the current position of sea stack Karl. This cone is

Figure 2.7 **(a)** Pillow lava at Valahnúkur; **(b)** lapilli tuff beds in the younger Vatnsfell tuff cone; **(c, opposite)** cross section showing the arrangement of the Vatnsfell tuff cones and Stampar feeder dykes at the coast of Reykjanes.

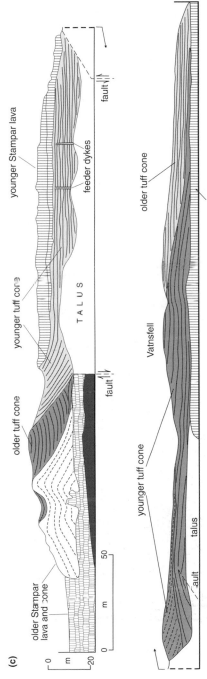

(c)

older Stampar
lava and cone

older tuff cone

younger tuff cone

younger Stampar lava

feeder dykes

T A L U S

fault

fault

0   m   20

0   m   50

younger tuff cone

Vatnsfell

talus

fault

older tuff cone

Skálafell lava

probably the mountain, described in the thirteenth-century chronicles, that rose out of the sea of Reykjanes in 1211. The lavas of Yngri Stampar were formed either in the same eruption or the second eruption of the Reykjanes Fires in 1223. Some of the feeder dykes dissect the Vatnsfell tuff connecting to the overlying flows of the Yngri Stampar cone row.

The thirteenth-century chronicles also mention an explosive eruption of the shore of Reykjanes in 1226, which caused widespread tephra to fall on land. This tephra layer is known as the Middle Ages Tephra (Miðaldar-lagið) and is found in soils all over the Reykjanes Peninsula, including Reykjavík and the Esja region. It forms an important time marker and has been used extensively for dating historical lava flows in the area. Eldey, a steep-sided islet 14 km off shore, is yet another remnant of an emergent submarine volcano formed by the Reykjanes Fires and it may be the source vent for the Middle Ages Tephra. Now it hosts the largest gannet-breeding ground in the Northern Hemisphere and its top is completely covered by a knee-deep layer of guano.

The chronicles also mention eruptions at Reykjanes in 1231, 1238 and 1240. The last eruption made the Sun appear "red as blood", which implies volcanic plumes rich in sulphuric aerosols, and atmospheric perturbations on a regional scale. Such plumes can be generated by effusive basalt erup-tions, provided that the volume of erupted magma is reasonably large. The only historical eruptions that fit the requirements of correct setting, age and size are those that produced the Arnarsetur and Illahraun lavas north of Svartshengi, some 18 km to the northeast of Reykjanes (see Fig. 2.1). Tephrochronology shows that these two lava flows were formed several years after the Middle Ages Tephra fell in 1226 and thus may represent the final episode in the Reykjanes Fires.

The second-youngest volcanic formations at Reykjanes are the Eldri Stampar and Tjaldastaðargjá cone rows and lavas formed by another vol-canotectonic episode 1500–2000 years ago. Other fissure volcanoes in the area are of unknown age but definitely pre-date the volcanoes mentioned above, illustrating that fires have raged at Reykjanes periodically through-out the Holocene. The oldest Holocene formations at Reykjanes are the lava shields Sandfellshæð, Skálafell and Háleyjarbunga, which together are the largest lava formations in the area, solely composed of pahoehoe flows. Háleyjarbunga may hold a special interest for petrology enthusi-asts, because it is composed of picrite basalt, which is exceptionally rich in large olivine phenocrysts. Some of the lava outcropping in the walls of the summit crater and the cliffs along the southern coast is green with olivine.

On the way. Continue on Road 425 along the southern cost towards the town of Grindavík. The route will take you through several older Holo-cene lava flows and, as you approach the putting greens of the Grindavík golf course, the mountain Þorbjarnarfell will appear on the skyline to your left. It is the highest (243 m) subglacial móberg volcano within the Reykjanes volcanic system and is noteworthy for the spectacular graben structure that cuts across its top. On the other side of Þorbjarnarfell is Svartshengi, the geothermal field that supplies the communities on the Reykjanes Peninsula with hot water for domestic use and house heating, but probably better known for its spa, the Blue Lagoon. It is an ideal spot for a relaxing dip in the warm mineral-rich water to recharge the batteries for the second half of the excursion.

Return to Road 245 travelling east towards Krýsuvík, which initially traverses the lavas of Sundahnúkar and Vatnsheiði. The former lava flowed into the sea to form the point Þórkötlunes and the inlet that hosts the harbour at Grindavík. It is one of the larger Holocene fissure-fed flows on Reykjanes (volume 0.6 km³) and it poured out of a 8.5 km-long fissure about 2400 years ago. The olivine-rich Vatnsheiði lavas are pahoehoe flows derived from three small picrite lava shields northeast of Grindavík that were constructed by eruptions during the early Holocene. The Vatnsheiði lavas host abundant gabbro xenoliths and large (2–3 cm) feldspar xenocrysts, which are readily discernible in the coastal outcrop in front of the table mountains Húsfjall and Fiskidalsfjall (see Fig. 2.1: site 9).

A little farther to the east, where the road climbs a steep incline, is Festarfjall, at 100 m high the tallest seacliff on the Reykjanes Peninsula, which consists of weakly consolidated hydroclastic tephra and is capped by a lava flow. The feeder conduit that fed the summit lava can be seen as a dyke dissecting the tuff sequence in the cliff facing the sea. The mountain's name is derived from the dyke, which was thought by locals to be the chain anchoring the mountain to the bedrock; hence, the name Festarfjall ("anchored mountain"). In geological terms, Festarfjall is a remnant of a submarine table mountain, a Surtseyan volcano that was formed in an eruption towards the very end of the Weichselian glaciation. A word of warning: rockfall is a common occurrence at Festarfjall and therefore we do not recommend hiking along the sandy beach in front of the cliff face. Northeast of Festarfjall is the 350 m-high table mountain, Fagradalsfjall, with a cap of subaerial lava indicating that the Weichselian glacier was at least 300 m thick in the area when the volcano was formed.

Locality 2.6 The Ögmundur lava
In passing the southern end of the móberg ridge Vesturháls, one enters the rugged lava flowfield of Ögmundarhraun, renowned for the ruins of the old Krýsuvík homestead laid to waste by the lava in the twelfth century. After the eruption, the homestead was rebuilt farther to the east at Bæjarfell. It is also one of the few lava flows in Iceland that takes it name from

a person. The folklore tells the story of Ögmundur "the Berserk", who cleared a passageway across the lava for the squire at the Krýsuvík estate, and upon completion of the "road construction" he was to be rewarded by marriage to the squire's daughter. However, the squire had no intention of giving his daughter away to the peasant and had Ögmundur killed just as he was about to complete the job. Since then the lava has been known as Ögmundarhraun.

Ögmundarhraun is one of several lava flows formed in the volcano-tectonic episode, known as the Krýsuvík Fires, that raged within the Trölladyngja volcanic system between 1151 and 1188. The lava flows, including Ögmundarhraun in the south and Kapelluhraun in the north, cover 36 km^2 and were produced by a series of volcanic fissures trending northeast for 28 km across the peninsula (see Fig. 2.1).

The old homestead of Krýsuvík was apparently a prosperous estate in its time, with a church and also rich fishing grounds off shore. The ruins, including the homestead, the church, and two sheep pens, are clearly visible in the kipukas Húshólmi and Óbrennishólmi, and can be accessed by walking down along the eastern margin of Ögmundarhraun and then following the walk path across the lava flow to Húshólmi.

Locality 2.7 The Krýsuvík area
Continue onwards to the east and then turn north on Road 42 towards Kleifarvatn. About 1 km up the road is the Grænavatn, the largest explosion crater of a small maar volcano complex that features eight craters in total. The Grænavatn maar is about 300 m in diameter and 44 m deep. The stratigraphy in the inner wall of the maar shows that the eruption began with powerful explosions ejecting tephra and fragments of the country rock. The event culminated with an effusive eruption, producing the small lava flow that caps the tephra sequence. The Grænavatn maar is well known for olivine-gabbro xenoliths, and their occurrence in the tephra deposits suggests that the activity resulted from deep-rooted explosions, perhaps as deep as 2–3 km.

Lake Kleifarvatn (9.7 km^2) is one of the deepest lakes in Iceland (107 m) reaching below sea level. It sits in a fault-bounded basin between móberg ridges Sveifluháls and Vatnshlíð and, like many other lakes in Iceland, it has no surface runoff. The móberg ridges on either side of Kleifarvatn were built up by a series of effusive and explosive subglacial eruptions, as is evident from the pillow lavas and tuffs that appear in outcrops along the road. At the foothills of Sveifluháls are the hot springs and the old sulphur mine at Krýsuvík. Return to Reykjavík by continuing north on Road 42, which intersects Highway 41 at the southern border of Hafnarfjörður.

Chapter 3

The south

General overview

South Iceland is bounded on either side by the active West and East Volcanic Zones (Fig. 3.1a). The South Iceland Seismic Zone transects the lowlands from the Hekla volcanic system in the east to the Brennisteins-fjöll volcanic system in the west (see Fig. 1.2). It is the source of some of the largest earthquakes in Iceland, with events between 6 and 8 on the Richter scale occurring periodically every hundred years or so (Fig. 3.1b). At the time of writing, the most recent major event occurred in the summer of the year 2000. At the surface, the South Iceland Seismic Zone is characterized by north–south trending strike-slip faults, but is thought to represent a major east–west trending transform fault linking the two volcanic zones.

South Iceland is characterized by contrasting topography, where the foreground is the vast flat-lying southern lowlands that are surrounded on the periphery by rugged mountains rising sharply up to 1000–1500 m. During glacial times the weight of the ice pushed the land down, so that in places it was well below sea level. When the Ice Age (Weichselian) glacier retreated from South Iceland about 10 000 years ago, the lowlands were temporarily inundated by the sea and became the bottom of the then vast South Iceland Bay (Fig. 3.2). During the early Holocene the land gradually emerged from the sea free of the weight of glacial ice, and vast sandur plains developed as rivers emerging from the interior highlands dumped their sediment load onto the flat coastal plains. The Great Þjórsá lava flow that now covers the area between the Þjórsá and Hvítá-Ölfusá River made a significant contribution to this process some 8500 years ago (Fig. 3.1a).

The Great Þjórsá lava is the oldest flow of the Tungnaár lavas, a series of eight lava flows that emanated from fissures within the Veiðivötn volcanic system to the north of Landmannalaugar during the Holocene and flowed to the southwest across the highlands and onto the Southern Lowlands. It is the largest of many voluminous fissure-fed basalt-lava flows that were erupted in Iceland at the beginning of Holocene, with a total volume of about 22 km^3. The largest lava-shield volcanoes are of similar

Figure 3.1 South Iceland. **(a)** The main geological features and excursion routes, stops and sites mentioned in the text; **(b)** occurrence and extent of damage by earthquakes on the South Iceland Seismic Zone.

size (e.g. Skjaldbreiður, 17 km³). The occurrence of these large-volume lava flows indicates that the production rate of basaltic magma at the beginning of the Holocene exceeds present-day production by a factor of 30–40. This high production rate coincides with rapid crustal rebound after removal of the glacier, up to 2000 m thick, of the most recent glaciation.

The mountains on the periphery of the region are either currently active central volcanoes or volcanic landforms formed by subglacial eruptions during the Pleistocene. The view from Oseyrarbru ("Óseyri Bridge") is one of the most breathtaking sights in Iceland, revealing a chain of towering central volcanoes from south to north: Eyjafjallajökull, Mýrdalsjökull, Tindfjöll, Torfajökull and Hekla. In the background to the north is a glimpse of Langjökull and Hofsjökull, the second and third largest glaciers in Iceland, the latter also being the largest active volcano in the country.

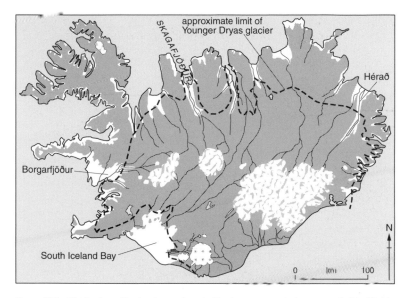

Figure 3.2 Distribution of lowland areas invaded by the sea towards the very end of the Weichselian glaciation. The extent of the Younger Dryas glacier is indicated by the dashed line.

We present two excursions in South Iceland. The first is sometimes nicknamed "the golden triangle" because it is centred on Þingvellir, the original site of the Icelandic Alþing ("parliament"), the Great Geysir – the grandfather of all erupting hot springs – and Gullfoss ("golden waterfall"), Iceland's most celebrated waterfall. The second excursion is to the Hreppar region, including Hekla, the queen of Icelandic volcanoes, and Þjórsárdalur, the region laid to waste by the first historical eruption at Hekla in 1104.

Þingvellir–Geysir–Gullfoss

On the route from Reykjavík to Þingvellir, the first part of the trip takes you through the older and deeply eroded Pliocene to Pleistocene succession, outcropping in the hills Úlfarsfell and Helgafell and, farther away, in Mt Esja (see Ch. 2). Along the road is the 17 km-long hot-water pipeline running from the geothermal area at Reykir in Mosfellsbær to Reykjavík, which has provided the city dwellers with thermal water for space heating since 1933.

As the road turns inland at the bottom of Leiruvogur and climbs up onto the Mosfellsheiði moor, the gently sloping Borgarhólar lava shield, one of the source volcanoes for the Grey Basalt foundation in Reykjavík, comes in to view towards the southeast. Another lava shield, Hædir, lu nn the eastern horizon. At the foothills of Mt Esja to the north of the road, the large gabbro body outcrops, representing the roots of the Stardalur central volcano. The volcano also produced the Móskarðshnjúkar rhyolite that caps the Mt Esja sequence in this area and shines brightly if there is a break in the cloud cover.

Locality 3.1 Þingvellir: the heart of Icelandic culture and heritage

No trip to Iceland is complete without a visit to the national park at Þingvellir. This historic site offers not only superb geology but also a glimpse into Iceland's history. Þingvellir is the original site of the Alþing, the general assembly of the Icelandic commonwealth (930–1262). It was established in 930 and is still operation today, making it Europe's oldest operating parliament. In the days of the commonwealth, each year the chieftains and their followers gathered at Þingvellir to settle their debates. The laws of the land were recited by the law speaker from Lögberg – a natural pedestal made of lava – using the walls of Almannagjá to echo his words over the constituency.

Þingvellir is truly the heart of Icelandic culture and it has been the venue for some of the most significant events in the country's history, perhaps none as significant as the nationwide adoption of Christianity in the year 1000, which came about through political debate at the Alþing. The main reason appears to be that a majority wanted to maintain peace and unity throughout the nation and to prevent civil war between the extremists within opposing factions, the followers of the old heathen religion and Christians. The records of the event also offer a rare indication of the contemporary understanding of volcanic eruptions.

As the members of the Alþing were debating the adoption of Christianity, news was brought of an eruption in the Ölfus district, South Iceland. It was apparent that the lava flow would overrun the farm of the heathen priest Þóroddur, and the extremists among the heathen followers spoke: "We are not amazed that the heathen gods are enraged at such a decision." Then the heathen chieftain Snorri replied: "At what were the gods enraged when the lava on which we are now standing was formed?" According to the chronicles, this reply was the turning point in the debate and ended the protest by the heathen extremists. It also shows that volcanic eruptions were generally viewed as phenomena of nature rather than punishment from higher authorities. For further information on the other historical aspects of Þingvellir, visit the information centre at the park entrance.

Viewing the Þingvellir area from the main lookout (see Fig. 3.1 · site 1), it is immediately clear that the geology of the area is one of seafloor spreading, displaying the intricate association of crustal rifting and volcanism

within the West Volcanic Zone. The bedrock in the area consists of Holocene basalt lavas covering the central part of the fault-bounded lake basin,
known as the Þingvellir graben, and Pleistocene pillow lavas and hyaloclastites, forming the móberg ridges and table mountains aligned along
the periphery of the basin. The Þingvellir graben is completely circum-
scribed on all sides by volcanoes that belong to four active volcanic systems, the Prestahnúkur and Hrafnabjörg systems to the north and the
Hengill and Hrómundartindur systems to the south (see Fig. 1.5).

Starting in the north, the first in view is the strongly faulted Ármannsfell mountain, which was formed by a subglacial eruption that broke
through the Weichselian glacier to form a table mountain composed of
strongly olivine phyric basalt. Behind Ármannsfell to the right is Skjaldbreiður, the prototype of monogenetic lava shields. Exhibiting a perfect
shield shape, the volcano has a basal diameter of 16–18 km and stretches
right across the northern end of the Þingvellir graben. It is characterized
by gently rising (1–8°) outer slopes climbing from 350 m at the base to
1000 m at the summit. The Skjaldbreiður volcano was constructed by a
long-lived effusive eruption that probably lasted for 50–100 years and its
lavas have a total volume of 17 km^3.

To the east on the edge of the graben are the subglacial volcanoes Hrafnabjörg and Tindaskagi, displaying the classical forms of a table mountain
and a móberg ridge, respectively. Farther east in the background are the
mountains of the Laugarvatn region, where J. G. Jones (1969, 1970) established the characteristic facies associations for subglacial volcanoes in his
classic studies during the late 1960s.

The extensive pahoehoe lava flowfields that fill in the graben in front
of Skjaldbreiður and Ármannsfell were mainly formed by effusive eruptions within the Hrafnabjörg volcanic system in the early Holocene. One
is the eruption that constructed the Skjaldbreiður volcano, whereas the
other three originated from relatively short fissures to the east of Hrafnabjörg mountain. Three of these lavas were formed in a major volcanic
episode about 9000 years ago, issuing 30 km^3 of lava into the Þingvellir
graben.

The first to form was the Þingvellir lava (C^{14} age 9130 ± 260 years) that
now floors the northern half of Lake Þingvallavatn and the area immediately to the north and east of the lake, and originated from volcanic fissures
to the southeast of Hrafnabjörg. The flows of this lava are exposed in the
fault scarp at Almannagjá, where they exhibit the classic structure of
inflated pahoehoe lobes. Model calculations show that it took more than
a year to accumulate the vertical thickness of the succession exposed at
Almannagjá. Every level in the succession consists of a series of laterally
arranged sheet-like lava lobes, each formed by successive breakouts from

75

an earlier lobe, so it is obvious that the Þingvellir lava was formed by a long-lived eruption that may have lasted for decades.

This activity was closely followed by the Skjaldbreiður lava-shield eruption, which, in addition to constructing the lava shield in the northern part of the graben, sent lavas southwards, which partly covered the existing Þingvellir lava. This episode culminated with the Eldborgir fissure eruption, which produced the rubbly-surface flows that cover the eastern part of the Þingvellir lava. These eruptions had a major impact on the evolution of Lake Þingvallavatn, because they both reduced its size and changed its hydrological character (see p. 77). The youngest lava, Þjófa-hraun, was formed by a fissure eruption some 2000–3000 years later.

To the east, the Þingvellir graben is bordered by the interglacial (Eemian) Lyngdalsheiði lava shield and closed off to the south by the jagged mountains that make up the Hrómundartindur and Hengill central volcanoes, which were mainly constructed by repeated subglacial centralized eruptions during the Weichselian glaciation (see Fig. 3.1). The irregular form of these central volcanoes is a good example of how the common conical shape can represent modified subglacial volcanism. The Hrómundartindur volcano has produced the most basic and the most evolved rocks in the Þingvellir area, namely the picritic pillow basalts of Mælifell and the andesites of Stapafell. Both central volcanoes are very much alive, as is clearly indicated by occurrence of Holocene eruptions and the many hot springs and fumaroles venting the high-temperature geothermal systems at Nesjavellir and Ölkelduháls.

The Tjarnahnúkur scoria cone, a parasitic vent on the Hrómundartindur volcano, erupted in the early Holocene (> 9000 years ago) to form small basaltic pahoehoe lava, which reached the shores of a proglacial lake that marks the embryonic stage of Lake Þingvallavatn. During the Holocene, four small fissure eruptions have occurred within the Hengill volcanic system, with vent sites on the north flanks of the volcano and within its fissure swarm. The youngest and most prominent was the 1900-year-old eruption that formed the Nesjahraun lava and the double-crater tuff cone of Sandey, the island in the middle of lake. The Sandey cone represents the hydro-magmatic phase of the Nesjahraun eruption, the only eruption known to have occurred within the boundaries of Lake Þingvallavatn.

The area's prominent feature is the spectacular northeast-trending Þingvellir graben nested within the northern limb of the Hengill fissure swarm. The graben, 10–20 km wide, narrows to the southwest, and is bounded by a series of normal faults on either side, the main ones being Almannagjá in the west and Hrafnagjá to the east. The graben floor slopes to the southwest, reaching 600 m altitude around the Skjaldbreiður lava shield and dropping below sea level at Lake Þingvallavatn.

Perhaps the most dramatic structure at Þingvellir is the giant crack of Almannagjá, where the ground has simply been ripped apart by the forces of seafloor spreading. Almannagjá is 7.7 km long, its greatest width 64 m and its maximum throw 30–40 m. The opposing boundary fault, Hrafnagjá, is 11 km long, 68 m at its widest, and has a maximum throw of 30 m. These and other boundary faults of the Þingvellir graben are thought be the surface expression of deep-rooted normal faults that extend through the crust, formed as the result of seafloor spreading. Within the Þingvellir graben are open fissures formed by the same process.

During the past 9000 years the graben floor has subsided some 40 m (averaging about 1 mm/year) between Almannagjá and Hrafnagjá, and the estimated horizontal extension is of the order of 70 m (averaging about 7 mm/year). Rifting within the graben is episodic, and individual tectonic events are associated with vertical and horizontal movements of the order of several metres. The most recent rifting episode occurred in 1789, when the graben floor subsided 1–3 m, with the consequence that the old parliament flats became unusable as a venue for the annual assembly of the Alþing. Since then the Alþing has resided in Reykjavík.

Þingvallavatn, the largest lake in Iceland, covers 82 km² and fills the deepest part of the graben. With a maximum depth of 114 m, it descends to 10 m below sea level. The present-day Þingvallavatn began to form about 9700 years ago as a proglacial lake in front of the retreating Weichselian glacier (Fig. 3.3). As the glacier retreated farther, rivers emanating from glaciers in the Langjökull area discharged into the basin and consequently the lake grew to a size approximately equal that of its present extent. The eruptions of the 9000-year-old Þingvellir–Skjaldbreiður–Eldborgir lavas had a profound impact on the hydrological character of the Þingvellir graben because they effectively terminated the surface flow of glacial rivers from the north and reduced the lake size by more than 50 per cent (Fig. 3.3). Ever since, the glacial meltwater discharging from the southern end of Langjökull has percolated through the porous lava formations, undergoing a natural filtering in the process, before discharging as freshwater springs into the lake. As a result of this steady groundwater-driven discharge and continuing subsidence of the graben floor over the past 9000 years, the lake as gradually grown to reach its current size.

To reach the next locality, follow Road 36, which takes you across Hrafnagjá, the fault that marks the eastern margins of the Þingvellir graben. Then turn at Gjábakki and follow Road 365 over the Lyngdalsheiði interglacial lava shield. The route across Lyngdalsheiði provides good views of the internal architecture of the subglacial volcanoes of the Laugarvatn Formation, with pillow lava and the capping-tuff sequence of the Kálfstindar móberg ridge appearing first, followed by the lava-crowned table

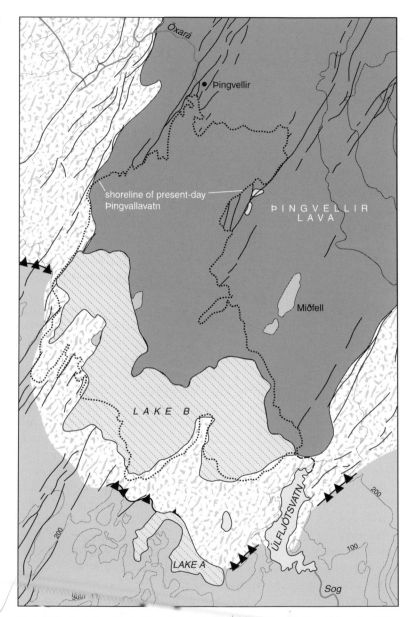

Figure 3.3 Evolution of Lake Þingvallavatn during the Holocene: lake A is the proglacial lake stage formed as the late Weichselian glacier retreated from its maximum stand in early Pre-boreal times; lake B inferred extent of the lake at about 9000 years ago when the northern part of the Þingvellir graben was flooded by the voluminous Þingvellir–Skjaldbreiður–Eldborgir lava flows, changing it to a spring-fed lake. Over the past 9000 years the lake has gradually enlarged because of consistent subsidence within the graben.

mountain of Laugarvatnsfjall. At Laugarvatn, geothermal springs occur along the northern shore of the lake and are utilized for heating natural saunas, which provide the ultimate perspiration experience. From Laugarvatn follow Road 37/35 eastwards to the Geysir geothermal field.

Locality 3.2 Geysir geothermal field

Geysir is perhaps the world's most famous spouting hot springs and its reputation is such that in many languages such springs are known as geysers. Geysir is situated among other hot springs in a field of silica sinter deposit at the eastern end of the hill Laugarfjall. Most of the springs here are alkaline with clear water. Geysir is by far the largest spring and it consists of a 3 m-wide and 22 m-deep shaft that opens into a 2 m-deep and 15 m-wide bowl-shape basin at the summit of the regular silica sinter dome that surrounds the hot spring. The hot springs are first mentioned in the Geysir area after the South Iceland earthquake in 1294. It is possible that the Great Geysir was rejuvenated by this event, but the 2800-year-old Hekla tephra layer that sits immediately beneath the silica sinter dome indicates a considerably longer lifetime for Geysir.

Geysir eruptions are initiated by superheating of the water column at about 10 m depth in the shaft, and the onset of an eruption is usually indicated by rumbling in the ground, vigorous boiling and periodic inflation of the water in the bowl forming a large cupola-shape bubble. As the bubble bursts, the eruption begins by sending up a series of water jets, growing in height until the event has reached its maximum. In its prime, Geysir produced 30–80 m-high jets in eruptions that lasted 5–10 minutes. In 1750 the Icelandic naturalists, Eggert Ólafsson and Bjarni Pálsson, described a Geysir eruption as follows:

We stayed by Geysir in case it would erupt. Earlier we had thrown several boulders of silica sinter, which lay astray, down into the bowl of Geysir. At first we heard a heavy thud beneath us. It resembled the sound of canon fire from a distance. In all, the thuds were five and of increasing intensity, and at the same time the ground jolted, like it was going to be lifted or ripped open. Following the sixth thud, the first jet emerged from Geysir. Thereafter the intensity of the eruption increased with each thud and the water was ejected as columns. The boulders we had thrown into the bowl had split into smaller fragments. The largest of these fragments were close to fist size. These fragments were ejected with the water and some reached heights greater than their associated water columns. We located ourselves up wind from Geysir so we could best observe the eruption because the steam blocked the view on the downwind side. This was also done to avoid injuries and burns from the hot water falling down from the columns. From the beginning of this eruption we noticed that, not only was the water lifted into the air by each emerging jet, but also simultaneously the total volume of water retained within the crater was inflated and overflowed the banks of the bowl, mostly to the north. . . . This eruption was equivalent to the highest and greatest eruptions known at Geysir. Although it did not reach the height of Laugafell, a small hill, 70 fathoms [116 m] high, in the neighbourhood of Geysir. In this particular eruption the water columns rose to approximately 60 fathoms [100 m]. According to the inhabitants of Haukadalur, the height of the water column in some Geysir eruptions reaches heights equal to that of Laugafell, but eruptions of such intensity occur only before the arrival of heavy rainstorms. The eruption lasted for ten minutes, with jets ejected every three seconds, and the eruption featured around 200 jets in total.

There are many other hot springs in the Geysir area, such as Strokkur ("the churn"), which has been known to have erupted jets up to 100 m high (see Fig. 1.18), Smiður ("the smith"), Óþerrishola ("the drizzly hole"), and Sóði ("the slob"). The most beautiful spring in the area is without doubt Blesi, which consists of two holes, one with bluish and the other greenish crystal-clear water; this effect is caused by colloids in the water.

On the way. Heading east from the Geysir area on Road 35 takes you across cultivated flat-lying farmlands and then up along the western edge of the Hvítá gorge and straight to the Gullfoss waterfall.

Locality 3.3 Gullfoss and Hvítárgljúfur
Where the glacial River Hvítá descends from the highlands onto the Southern Lowlands, it has eroded an impressive gorge into the edge of the highlands, along faults trending north-northeast. The Hvítá gorge (Hvítárgljúfur) is 2500 m long and its maximum depth is 70 m. The two-step waterfall of Gullfoss is at the head of the gorge and has a total drop of 32 m. Although the course of the river is controlled by faults, Gullfoss and the gorge owe their existence to the contrasting lithologies that make up the succession in the Gullfoss area. The succession is characterized by fairly thick sedimentary units, each capped by somewhat thinner lava flows (Fig. 3.4). The lavas are resistant to erosion and represent the formations that carry the waterfall. The sedimentary units are softer than the lavas and are therefore eroded faster by the water. This results in undercutting of the lavas, forming an overhang, which eventually collapses and the process is repeated. This step-wise process has carved the gorge into the landscape in less than 10000 years, at an average rate of 25 cm/year.

Figure 3.4   (a) Gullfoss waterfall. (b) Graphic log of its stratigraphy: lower (B_1) and upper (B_2) threshold layers; lower (S_1) and upper (S_2) sedimentary units

From Gullfoss follow Road 35 back to Geysir and continue all the way to locality 3.4 at Grímsnes. The journey takes you across cultivated lowlands, passing by the low mountains Vörðufell and Hestfjall. Vörðufell is an eroded remnant of the Pliocene–Pleistocene basement rocks projecting through the early Holocene sedimentary cover. Several major strike-slip faults formed in association with earthquake activity in the South Iceland Seismic Zone dissect the mountain and are easily detected from a distance. Hestfjall is a subglacial volcano of the table-mountain type, formed by off-rift volcanism where magma has found a pathway to the surface through faults associated with the South Iceland Seismic Zone.

Locality 3.4 Grímsnes volcanic field
The Grímsnes volcanic field is an inconspicuous group of 12 scoria-cone volcanoes sitting about 15–20 km to the east of the main axis of the West Volcanic Zone and in line with the South Iceland Seismic Zone. The Gríms-nes volcanoes were formed in a series of mixed eruptions that took place 5000–8000 years ago and in total produced 1.3 km³ of lava. Each volcano

81

is typically made up of a row of two to four scoria cones delineating short (300–1000 m) volcanic fissures and a small lava flowfield 3.5–23.5 km^2 in area. The exception is the Kerhóll volcano, where a circular spatter cone was formed around the erupting vent some 6200 years ago.

The largest edifice of the Grímsnes group is the Seyðishólar scoria cone volcano, featuring >100 m-high cones, whereas the oldest ones are the Selhóll cones. The best-known edifice is undoubtedly the crater Kerið, a popular tourist stop that it is often mistakenly cited as an example of an explosion crater or a maar volcano because of the small pool retained within it. Kerið is a mixed cone made up of spatter and scoria, which belongs to the Tjarnarhólar cone row. In the eruption some 6300 years ago, Kerið contained a small lava pond that was drained in the final stages of the eruption, leaving behind a deep and steep-sided crater. The crater was deep enough to penetrate the local groundwater table, allowing water to flow freely into the crater to form the small lake.

On the way. In crossing the River Sog, the road turns along the foothills of Mt Ingólfsfjall, a table mountain that rests on a pedestal of a series of significantly older lava flows that belong to the Hreppar Formation. The large boulders and blocks that occur in abundance at the base of the mountain are a stark reminder of the forces that loom beneath your feet, because most of them have fallen from the upper cliffs during major earthquakes in the South Iceland Seismic Zone.

Continuing westwards on Highway 1, the large scarp that appears on approaching the town of Hveragerði is an old seacliff carved out into young volcanic formations by the invading seas at the end of the most recent glaciation. As the winding road climbs up this old seacliff, we re-enter the West Volcanic Zone. On a good day the view from top of Hellisheiði is a gem. To the east are the Southern Lowlands, with the volcanoes of the East Volcanic Zone, including the Westman Islands, in the background. To the west, Reykjavík appears in the distance, with the Hengill central volcano standing tall in the foreground. Upon descent, the highway crosses a young-looking lava flow, sometimes referred to as Christianity lava (also known as Svínahraun), which is generally believed to be the lava formed by the eruption mentioned when the members of the Alþing were debating the adoption of Christianity in the year 1000 (see p. 74).

Hreppar–Þjórsárdalur–Hekla

This excursion is designed such that one can proceed directly from the Gullfoss–Geysir area. From Geysir follow Road 30 to the south towards the town of Flúðir and continue to the intersection with Road 32 and then turn left (signposted Árnes, Þjórsárdalur and Sprengisandur). This is the starting point of the second excursion.

Here we are practically on the shore of the ancient South Iceland Bay about 100 m above present sea level. Before us are the Southern Lowlands, which were at the bottom of the sea towards the end of the Weichselian glaciation (see Fig. 3.2). Shortly afterwards, the lowlands emerged from of the sea because of the coupled effects of isostatic rise of the land and the construction of vast sandur plains through the accumulation of debris transported by glacial rivers. About 7 km to the east at the Búðafoss ("Búði rapids") are the outermost moraines formed during the Younger Dryas stage, where the glacier was calving into the sea some 10 000–11 000 years ago.

On the way. The road from Árnes to Þjórsárdalur runs parallel to the River Þjórsá, the longest river in Iceland, stretching some 230 km from its source at Hofsjökull to the southern coast, and which has an average discharge of 400 m³/s. Along this stretch are islets in the middle of river, which are covered by lush vegetation, because their location prevents intervention by man or livestock. South of the river is the Great Þjórsá lava (see p. 71). The mountains and hills north of the road belong to the Hreppar Formation, a Plio-Pleistocene sequence built mainly of subaerial basaltic lava flows and clastic deposits of fluvial and debris-flow origin. Several distinctive tillite horizons and thick glacial lagoon deposits occur within the upper part of the sequence, indicating the onset of full-scale glaciation about 2.4 million years ago. The Hreppar succession is heavily dissected by faults, and the tectonic fabric is characterized by older north-trending strike-slip faults and younger northeast-trending normal faults.

Locality 3.5 Gaukshöfði–Villingadalsfjall sandur-plain succession
Pull off onto the side road by Gaukshöfði. The southern slope of Mt Villingadalsfjall may at first glance look like any other mountain slope in the region and certainly its geological significance is not revealed in an instant. The lower part of the slope features large grass-covered terrace-like benches, whereas barren rock is exposed in the steeper upper part. The benches are in fact large blocks that have been displaced downwards by faulting. For example, the rock formations exposed in the cliffs by the main road at Gaukshöfði, which includes an excellent outcrop of pillow lava, are the same formations as occur at the top of Mt Villingadalsfjall, 300 m up.

Walking up the lower half of the slope, these formations are repeated at each crossing of a northeast-trending ravine. The ravines are in fact the fault traces. Occasionally a basaltic dyke pokes up in the ravines; the dykes used the fractures created by the faults to rise towards the surface. The

Figure 3.5 Simplified cross section of the geology of Villingadalsfjall.

benches were formed by a step-wise displacement of the strata along north-east-trending faults and, in all, the cumulative downthrow is nearly 300 m. The faults mark the northern margins of the East Volcanic Zone.

In the upper half of the slope, the original stratigraphy is intact and it reveals an interesting story. The sequence exhibits distinctive stratification and most of it is composed of fluvial conglomerates and sandstones (Fig. 3.5). Scattered in between these fluvial deposits are lenticular lava bodies with upper surfaces that show clear evidence of having been modified by fluvial erosion. The fluvial deposits were formed by a braided river system emerging from a glacier situated farther inland. In fact it is an old sandur plain, where the debris transported by the ancient rivers and sporadic lava eruption filled in a large palaeo-valley. The western slope of this valley was the east side of Mt Hagafjall, where the móberg ridge that forms the mountain core is blanketed by steeply dipping alluvial-fan deposits.

Locality 3.6 Þjórsárdalur
Þjórsárdalur Valley is cut into an extinct central volcano, which bears the same name. This is revealed by an unusual abundance of andesite and rhyolite dyke, lava and tephra formations in the Plio-Pleistocene strata of the surrounding mountains. The valley is in fact an old caldera that has been extensively modified by glacial erosion. The volcano centre is at the head of the valley near Mt Fossalda, where many inclined cone dykes dissect strongly altered and colourful rhyolite and andesite formations.

Sections of the original caldera faults are seen at Mt Skeljafell on the east side of the valley and at Grjótárgljúfur on the west side, showing that the caldera was at least 8 km wide, rim to rim. At these localities, rhyolite dykes have intruded the caldera faults and connected to small rhyolite lava domes a little higher up in the sequence. At Mt Skeljafell (see Fig. 3.1: site 2) the dyke is exposed in a small ravine north of the main road, where it ascends the western slope of the mountain and can be followed up to the lava dome (the pale rock formation) where it sits on the caldera fault.

Among the rocks that form the caldera fill are andesite tuffs and pillow lavas that are closely associated with tillite beds. A good example of this can be seen at Mt Reykholt in the centre of the valley, where a thick pillow lava and móberg tuff sequence is sandwiched between two tillite layers. This sequence clearly shows that, about 2 million years ago, while the volcano was still active, a glacier filled in the caldera. Later the volcano was almost completely buried by a series of basaltic lava flows formed by eruptions outside the Þjórsárdalur central volcano, and then its caldera was carved out by erosion during subsequent glaciations forming the valley we see today.

In the past 7000 years Þjórsárdalur has been repeatedly devastated by tephra fall from the nearby Hekla volcano, but never more so than by the largest Plinian eruption at Hekla some 2900 years ago. The whitish specks on the surrounding mountain slopes are not snow but the remnants of the pumice-fall deposit from this Hekla eruption. At the foothills of Mt Skeljafell this fall deposit is more than 2.5 m thick. When the first settlers arrived in Iceland in the ninth and tenth centuries, the valley had completely recovered from its previous encounters with Hekla and it soon became quite densely populated, hosting at least 20 farms. This settlement in Þjórsárdalur Valley was decimated by the tephra fall from the 1104 eruption at Hekla. At the time, the valley was completely blanketed by a 10–30 cm-thick pumice-fall deposit, which was enough to make the area uninhabitable. Some of these ancient farm ruins were excavated by the Nordic archaeological expedition in 1939 and one of them is the old Norse homestead at Stöng (see Fig. 3.1: site 3).

The valley floor is covered by a 3500-year-old lava flow – one of the youngest Tungnaár lavas, which originated from volcanic fissures in the Veiðivötn area. It is a branch of a larger lava flowfield, the Búrfell Lava, which covers the high plateau to the east Mt Skeljafell and Mt Búrfell. The lava found its way into the Þjórsárdalur Valley through a narrow ravine named Gjáin, which is about 500 m to the east of the farm ruins of Stöng. Gjáin is a beautiful hollow with two small waterfalls surrounded by grassy slopes, which makes it an ideal place for a picnic on a good day. The geology at Gjáin shows that this small gully has been a feature of the landscape for a quite some time and that the Búrfell Lava is not the first one to flow through it (Fig. 3.6).

In the centre of the Þjórsárdalur Valley, the Búrfell Lava contains a cluster of small scoria cones, which were formed by rootless eruptions when the lava covered the wetlands that once existed in the valley. At Hjálparfoss the River Fossá has eroded a beautiful section through the lava, exposing the roots and conduits of the rootless cones, along with spectacular fanning columns formed by water-enhanced cooling of stagnant lava.

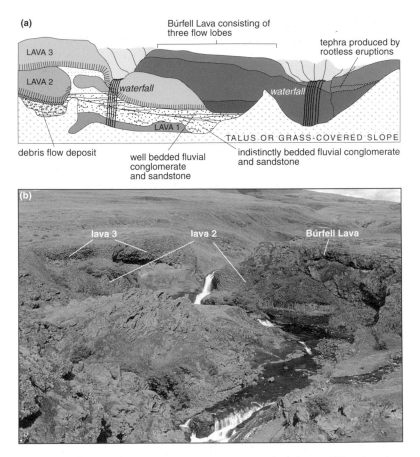

Figure 3.6 **(a)** Cross section showing the main geological formations at Gjáin, where four separate lava flows (lava 1, 2, 3 and Búrfell Lava) fill in pre-existing river channels; **(b)** a photograph of Gjáin that corresponds to the left half of the section.

On the way. Travelling from Þjórsárdalur to Hekla the route (Road 32) crosses a pass that the Tungnaár lavas followed down to the lowlands. The major glacial rivers, Þjórsá and Tungnaá, also flow through this pass, and a suite of dams have been built in this area to serve hydroelectric power plants that produce more than 75 per cent of the country's electrical energy. The route takes you by the Búrfell and Sultartangi hydroelectric power plants. As you cross the Sultartangi bridge, turn south on to Road 26. Turn east at the intersection with Landmannaleið and follow it to the destination at Skjólkvíar. At that point, one should have a good view of the queen, the Hekla volcano.

Locality 3.7 Hekla – the queen of Icelandic volcanoes
Hekla (1491 m) is without doubt Iceland's most famous volcano and is the
third most active volcano in the country, behind Grímsvötn and Katla,
with a tally of 18 eruptions in historical times (Fig. 3.7). Hekla is the only
one of its kind, because it is the only ridge-shape stratovolcano known in
Iceland where eruptions occur repeatedly on the same fissure (see also
p. 31). Hekla magmas are erupted from an 8 km-deep chamber that contains two magma types: an andesite magma containing 52 per cent SiO_2
(by weight) and a rhyolite magma with SiO_2 in excess of 70 per cent. Intermediate magma compositions, which are common, are produced by mixing of the end-member magma types in the chamber prior to eruption.

Hekla has produced a vast number of eruptions in postglacial time,
although the three rhyolitic Plinian eruptions H3 (2900 BC), H4 (4500 BC)
and H5 (6500 BC) stand out as they blanketed up to two-thirds of the country with pale pumice fall deposit (Fig. 3.7). The first eruption in historical

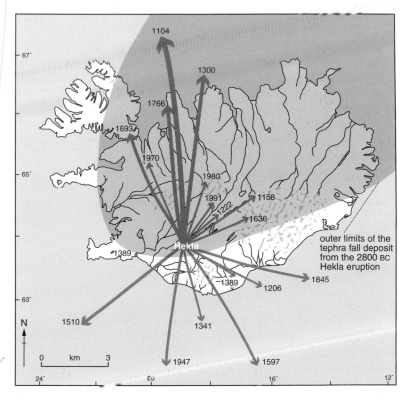

Figure 3.7 Tephra-dispersal direction for historical Hekla eruptions (and years of eruption);
the shaded area shows the extent of the H3 tephra layer formed 2900 years ago by a Plinian
eruption at Hekla.

87

times was tremendous explosive Plinian eruption of 1104, emitting 2.5 km³ of rhyolite tephra covering more than half of Iceland with pumice. More commonly, Hekla events are mixed eruptions that begin with a shortlived (<1 hour) explosive phase at the summit crater, producing Plinian fall deposit of dacite or andesite composition. Then there is a sharp transition into a fissure eruption, with a curtain of fire extending across the crest of the volcano producing andesite tephra and lava. This phase typically lasts between several hours and a couple days, when the activity becomes confined to a short segment of the fissure or a single vent that primarily produces lava.

Hekla erupted four times in the twentieth century – in 1947, 1970, 1980 and 1991 – and celebrated the turn of the millennium by an eruption on 26 February 2000. Records of other historical Hekla eruptions show that the volcano has erupted at least once every hundred years since 1104, except in the fifteenth century (Fig. 3.7). All of these eruptions are mentioned in the historical chronicles, thus facilitating more exact dating of the eruptions than would otherwise be possible. Some eruptions are described in great detail, especially the younger ones (i.e. post-1693), and demonstrate a surprisingly pragmatic attitude towards these fearsome events. This is clearly illustrated by the description of the eruption at Hekla in 1300, written by Einar Hafliðason in the fourteenth century:

Coming up of fire in Mount Hekla with such violence that the mountain split, so that it will be visible as long as Iceland is inhabited. In that fire, great rocks played like coals and when they hit together there were great crashes that were heard in the north of the country and in many places elsewhere. So much pumice flew onto the homestead at Næfurholt that the roofs of the buildings were burnt off. The wind was from the southeast, and it carried northwards over the country such dense sand between Vatnsskarð and Axarfjarðarheiði, along with such great darkness, that no-one inside or outside could tell whether it was night or day, while it rained the sand down on the earth and so covered all the ground with it. On the following day the sand was so blown about that in some places men could hardly find their way. On these two days people in the north did not dare put to sea on account of the darkness. This happened on 13 July.

The total volume of magma erupted by Hekla in historical times is about 10 km³, which is enough to build a 1 m wide and 1.6 m-high wall around the 6000 km-long coastline of Iceland. The tephra fall from Hekla eruptions has repeatedly caused great damage to pasture and cultivated farmlands, with more than 50 farms in the vicinity (<70 km) of the volcano damaged or destroyed in a single eruption. In more distant regions, the damage felt was different, where much of the livestock was killed by fluorine poisoning through the intake of ash particles while grazing in fields affected by ashfall. The greatest damage was caused by the eruptions of 1300, 1341, 1510, 1693 and 1766.

The composition of the Hekla magma system is roughly a linear function of the repose periods between eruptions. As the repose period increases, the tephra produced by the initial Plinian phase of each eruption becomes more evolved (i.e. contains more silica) and subsequently the explosivity of the initial phase increases. However, the effusive phases that follow always produce lavas of andesite composition (52–58% SiO_2). This compositional pattern in the eruption products indicates chemical stratification of the magma in the holding chamber beneath Hekla. Such chemical stratification is established as the magma fractionates through processes such as magma mixing, crystal fractionation and assimilation of the surrounding crustal rocks.

The shape of this reservoir, its location at the west margin of a propagating rift, and its association with a crustal weakness, all contribute to the high eruption frequency of Hekla. In this context it is interesting to note that eruptions in the immediate vicinity of Hekla, such as the 1554 Rauðubjallar (Pálsteinshraun) and 1913 Lambafit–Mundafell fissure eruptions, produced basalt-lava flows – magma composition less evolved and distinctly different from that erupted by Hekla summit eruptions. These basalt magmas are most probably derived from a deeper-seated reservoir situated below the magma chamber of Hekla proper.

On the way. The travel from the Þjórsá bridge to Seljalandsfoss at Eyjafjallajökull takes you across cultivated sandur plains constructed by the glacial rivers over the past 10000 years. The area is in essence the marshland of a large delta, which explains why most of the farmhouses stand on hilltops. The low-lying areas used to be covered with peat bogs and fens, but have been drained extensively for cultivation. This region was the stage of Brennu–Njálssaga, one of the best known of the Icelandic sagas.

In the northeast, the ragged points of the Tindfjöll volcano rise to 1460 m, which are all named after trolls and other mythological figures, such as Saxi, Ýmir and Sindri. A large explosive eruption occurred at the Tindfjöll volcano about 200 000 years ago. This eruption produced widespread tephra fall that is found in the ocean all around Iceland, and also a large ignimbrite sheet that spread southwards over the Þórsmörk area, where it is exposed in accessible outcrops.

Seljalandsfoss at Eyjafjallajökull is an impressive 65 m-high waterfall that cascades off an ancient wavecut cliff and is unique in the sense that it can be viewed from the back as well as the front. An easily accessible footpath passes behind the waterfall, a walk well worth the effort. The grass-covered slopes at Seljalandsfoss are a good vantage point for viewing the vast sandur plain of the Markarfljót River. The farmlands in front of Fljótshlíð and Eyjafjallajökull stand on a sandur plain that has been constructed in less than 10000 years by a 60km-long braided river system stretching from the Þórsmörk area to the sea. The upper part of the sandur plain fills in a 2–6 km-wide valley, whereas the broader lower section (i.e. the Landeyjar area) is the fanning delta. The main river is Markarfljót, which originates at Mt Rauðufossafjöll, to the east of Hekla. It flows south in a narrow gorge through the mountainous region between Tindfjöll and Mýrdalsjökull before it enters the sandur plain at Þórsmörk. On its 100 km journey to the coast, the Markarfljót grows steadily as it captures the drainage from its tributaries, which feed off the meltwater from the surrounding glaciers. The total drainage area of the Markarfljót River is 1070 km^2.

The discharge of the Markarfljót River is 80 m^3/sec on average, but it fluctuates wildly, commonly with about a tenfold increase in flow during spring thaw or at times of heavy rain. Before the river was confined to its present course by manmade barriers and walls, the frequent fluctuations in the discharge resulted in an unstable course and, with time, the river channels wandered across the plain. In historical times these frequent shifts in the river course have caused severe damage to the farmlands.

Chapter 4

Vestmannaeyjar
(the Westman Islands)

General overview

In about AD 870 the brothers Ingólfur and Hjörleifur sailed for Iceland with
the intention of settling there for good. During the first winter, Ingólfur
made camp at Ingólfshöfði; Hjörleifur made his winter camp at Hjörleifshöfði. In the following spring Hjörleifur and his men were killed by his
Irish slaves (often referred to by the Vikings as "Westmen"), which after
the event fled to a group of small islands off the south coast. When Ingólfur
found out about the fate of his brother, he sailed to the islands and killed
all the Westmen. Thenceforth, the archipelago was known as Vestmannaeyjar ("Westman Islands").

The Vestmannaeyjar volcanic system is located on the seaward extension of the East Volcanic Zone and is characterized by alkalic magmatism.
Approximately 80 volcanic edifices are known within the 900 km^2 covered
by the system, of which 18 rise above sea level (Fig. 4.1). Vestmannaeyjar
is a young volcanic system that came to life < 20 000 years ago. The islands
forming the archipelago are even younger, as they were all formed in the
past 10–20 000 years and most by emergent submarine eruptions similar to
the 1963–7 Surtsey eruption (see pp. 95–96). However, the largest island,
Heimaey, is a central volcano in the making and is constructed from tight
cluster of basaltic volcanoes that were produced by series of eruptions
over the past 12 000 years (Fig. 4.1). The eruption of more-evolved mugear-
ite to Hawaiite magma in the 1973 Eldfell eruptions confirms the status of
Heimaey as a juvenile central volcano. Consequently, the geology of
Heimaey reveals a more diverse style of volcanism than is found on the
other islands, including the products of subaerial basalt and intermediate
mixed eruptions, in addition to explosive hydromagmatic eruptions.
Observations made during the 1963–7 Surtsey and 1973 Eldfell eruptions
show that the main magma-holding reservoir of the Vestmannaeyjar
system is located at 25–30 km depth and that it supplies the system with

Figure 4.1 Heimaey and other islands of the Vestmannaeyjar archipelago.

basaltic magma. The magma erupted in the 1973 Eldfell eruption is derived from the Surtsey magma by fractionation that occurred in a shallow holding chamber over the ten years that separated the eruptions.

Heimaey

Heimaey, the largest and the only inhabited island of the Vestmannaeyjar archipelago, is 5.5 km long and 4 km wide, with a total area of 11 km². It has about 5000 inhabitants and is the home of the largest fishing fleet in Iceland. The island is made up of five principal geological formations (Fig. 4.1), which are described in some detail below.

Locality 4.1 Norðurklettar
Norðurklettar ("northern cliffs"), which consist of Dalfjall, Klif and Háin to the west, and Heimaklettur, Miðklettur and Ystiklettur to the east, are the oldest geological formations at Heimaey. These cliffs were formed in

a series of eruptions towards the end of the Weichselian glaciation, 10 000–12 000 years ago. The Norðurklettar Formation is made up of products from several eruptions, which at the time formed the two separate islands that were moulded into their present shape by wave erosion over a period of 4000 years.

Locality 4.2 Stórhöfði and Sæfell
Approximately 6000 years ago another eruption created the third island, Stórhöfði, the southernmost point of Heimaey. Stórhöfði is a typical submarine table mountain, where a small pahoehoe basalt-lava shield caps the hydromagmatic tephra foundation. The eruption of Stórhöfði was similar to that of Surtsey, characterized by hydromagmatic explosive activity while building itself out of the ocean and becoming effusive when the island had grown large enough to prevent sea water from accessing the vents. Stórhöfði lava shield is a classic example of a composite tube-fed pahoehoe lava, as is evident from the many lava tubes exposed in the lava. About 5400 years ago a submarine explosive eruption began just north of Stórhöfði and formed the tuff cone Sæfell, the largest volcanic structure on Heimaey, with a diameter of 1 km. The Sæfell tuff cone was a sizeable addition to the island of Stórhöfði, but despite its size the cone was not large enough to connect all of the three older islands into one landmass, something achieved by the next major eruption of the Heimaey volcano.

Locality 4.3 Helgafell
The Helgafell scoria volcano (227 m above sea level) in the centre of the island was formed in an eruption about 5000 years ago. This eruption also produced the pahoehoe lava flowfield that covers the western side of the island and in the process connected the Sæfell–Stórhöfði and the Dalfjall–Klif islands into a single landmass. The lava also closed the seaway south of the Norðurklettar, providing the right conditions for the formation of an isthmus, Eiðið, the natural breakwater between Klif and Heimaklettur. The island of Heimaey was born.

Locality 4.4 Eldfell
The most recent addition to the island is the scoria cone Eldfell ("mountain of fire") and its lava flow, which was formed in an eruption in 1973. On Tuesday 23 January, at 1.55 a.m., an eruption began from a 1.6 km-long fissure on the eastern side of Heimaey. The eruption came with little warning. A few mild shocks were felt from 10 p.m. the night before, the sharpest of them occurring at 1.40 a.m. On Monday night, 22 January, people in Heimaey went to bed as usual after a normal workday. The whole fishing fleet was in dock because of a severe storm.

Just before 2 a.m. there was a telephone call to the local police station. Information was received that an eruption had started a short distance from a farm, Kirkjubær, located at the eastern edge of town. The police officers on duty drove immediately to the area, where they saw a volcanic fissure 500 m to the east of Kirkjubær, extending from the harbour mouth in the north and to the sea south of Helgafell. A series of fire fountains rose up from the fissure, spaced so closely that they formed almost a continuous wall or curtain of fire. The eruption was soon reduced to one vent, which is now the main crater of the new volcano, Eldfell. From the beginning, lava flowed from the fissure down slope to the east and northeast, and into the sea.

The fire alarm was sounded in town, while fire and police cars patrolled the streets with sirens in order to wake people. Within about two hours most of the population was afoot. For reasons of safety the town council decided to evacuate the whole town immediately, apart from a few volunteers who remained to conduct necessary duties. There was a danger of the harbour mouth being sealed off by the lava if the fissure should lengthen to the north, and similarly the airport was in danger if the lava extended southwards.

Contact was made with the civil-defence force, the local airlines and the NATO Defence Force in Keflavík to provide air transport for hospitalized and elderly people. That night, 300 people, mostly the sick and aged, were airlifted to Reykjavík. The rest of the town's population rushed down to the harbour, having just had time to put on warm clothing and a few necessary belongings into a duffle bag. Thanks to the storm, there were 60–70 boats in dock. All boats were now quickly prepared for departure; the first left for Þorlákshöfn at 2.30 a.m., followed by a steady stream of fishing vessels packed with people. The whole operation went remarkably smoothly and without mishaps, thanks to the incisive response from local authorities, favourable weather conditions and a little luck.

In the first 12 hours the eruption spewed forth more than 30 million tonnes of tephra and lava, which were initially dispersed to the north and east. The eruption lasted just over five months to 3 July 1973. It covered most of the town with several metres of tephra, and about a quarter of the town was bulldozed flat and buried by the lava. The size of Heimaey was increased by 2.2 km^2 in the eruption and the lava flowfield covers an area of about 3.3 km^2. The total amount of magma erupted was 0.25 km^3, as compared to the 1.2 km^3 erupted by the 1963–7 Surtsey eruption.

The 1963–7 Surtsey eruption
An island emerging from the sea

Surtsey is a small volcanic island situated about 33 km off the central-south coast of Iceland and is the westernmost island of the Vestmannaeyjar archipelago. The island is the subaerial part of the larger Surtsey volcano, a 6 km-long east-northeast-trending submarine ridge that rises from a depth of 125 m and covers about 14 km^2 (Fig. 4.2). The prominent features on Surtsey are two abutting 140 m-high tuff cones and a small pahoehoe lava flowfield that caps the southern half of the island.

Surtsey is also the youngest of the Vestmannaeyjar, formed by a prolonged eruption that was first noticed on 14 November 1963 by fishermen attending their nets about 20 km southwest of Heimaey. The eruption intensified and gradually built an island that in its prime rose to 170 m above sea level. It was named Surtsey after Surtur, the fire-raising giant of Norse mythology who was to set fire to the Earth on Judgement Day. Surtsey rumbled and lava flowed for 3½ years until June 1967, when Surtur called it quits. The longest eruption in Iceland since settlement was over and, for the first time, Icelanders had witnessed the formation of a submarine table mountain.

The tuff cones, Surtur I and II (Fig. 4.2), were built by Surtseyan explosive eruptions over a period of five months between November 1963 and

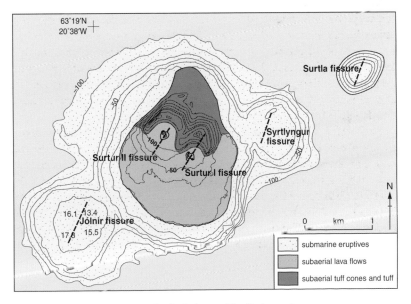

Figure 4.2 Geological map of the Surtsey volcano

95

April 1964. During this stage the eruption columns rose as high as 13 km. When the island had grown large enough to seal the volcanic conduit from the invading sea water, the activity changed from being wholly explosive to purely effusive. The eruption featured two prolonged subaerial effusive eruption phases, the first erupting through the Surtur II crater and lasting for 13½ months (April 1964 to May 1965). It produced a 100 m-high lava shield that covered 1.53 km² above sea level. In the second phase the Surtur I crater came to life again to produce a 70 m-high and 1 km-wide lava shield between August 1966 and June 1967. The total volume of tephra and lava produced by the Surtsey eruption amounts to about 1.2 km³, of which about 0.4 km³ is lava. Today, erosion is the main force at work on Surtsey and, since the end of the eruption in 1967, wave erosion has reduced the surface area of the island by 1 km².

Long before the eruption stopped, the island was proclaimed a nature reserve and all visits to it were restricted, so biologists could make the most of this unique opportunity to study how life would migrate to and develop on this new and desolate land. In 1964 the first living organism was found in the ash deposits closest to shore and in May that same year a single fly was found on the island. Seagulls hung out in the tidewaters farthest from the crater and the first vascular plant to flower on the island was the sea rocket, found along the shore in 1965. Today five species of birds nest on Surtsey: herring gull, black-backed gull, black guillemot, kittiwake and fulmar. The last species was the first one to nest and hatch its young successfully on Surtsey in 1970.

Chapter 5
The central south

General overview

Central-South Iceland extends from the Markarfljót River in the west to Jökulsá at Breiðamerkursandur in the east (Fig. 5.1). In the west it stretches across the East Volcanic Zone along the foothills of the glacier-capped Eyjafjallajökull and Mýrdalsjökull volcanoes, and then follows the margins of the zone to the northeast before turning east towards the Öræfajökull volcano across the vast and desolate plains of Skeiðarársandur. In the north the largest glacier in Europe, Vatnajökull, towers over the region. Currently, this is the most volcanically active area in Iceland, as can be deduced from district names such as Eldsveitirnar ("fire districts") and

Figure 5.1   The main geological features of Central-South Iceland.

Öræfi ("wasteland"). Over the past millennium it has experienced not only more but also larger eruptions than any other part of the country.

However, although destruction and hardship are brought on by the volcanic eruptions, the fertility of the soil is greatly enhanced by the tephra falls they produce, and the towering volcanic mountains generate clouds and precipitation. Consequently, the area enjoys the highest rainfall in Iceland and in between the rugged mountains are valleys with fertile pastures. This is the only region in Iceland where the land is painted green with vegetation from the sea to the highest mountaintops. Even the talus slopes are grass covered. As the soil cover creeps down the steep talus slopes, it often forms majestic stairways that in places appear to stretch up into the sky above.

We present two excursions across Central-South Iceland. The first covers the coastal regions from the Markarfljót River to Mýrdalssandur. The second is centred on the Fire Districts, the region between the large sandur plains of Mýrdalssandur and Skeiðarársandur, but also includes the regions around the national park at Skaftafell.

Eyjafjallajökull and Mýrdalur
The icecaps and the secluded volcanoes

This excursion starts at the Markarfljót bridge (Highway 1). It follows the foothills of Eyjafjallajökull to Skógar and then crosses the plains of Skóga-sandur and Sólheimasandur into Mýrdalur and on to the western part of Mýrdalssandur.

Eyjafjallajökull

Eyjafjallajökull is a 1666 m-high shield volcano, as is suggested by its gen-tly sloping east–west profile (see Table 1.4). The volcano is crowned by a small caldera and an icecap that reaches down to 1000 m above sea level. The south slope of the volcano terminates in old wavecut seacliffs that rise to 700 m above present sea level, whereas the steep north slope has been strongly modified by glacial erosion.

The Eyjafjallajökull volcano has been active for at least 800 000 years; it consists of six glacial and six interglacial volcanic sequences. The former sequence comprises subglacial volcanics, such as pillow lavas, kubbaberg breccia, and hydroclastic tuffs and intercalated tillites. The latter is typi-cally composed of a stack of lava flows indicating emplacement during ice-free periods. Good examples of these sequences are exposed in the sea-cliffs on the south side. The volcano has been relatively quiet in late glacial and Holocene times, with only two effusive basalt fissure eruptions, which subglacial lavas, kubbaberg and móberg breccia

thin lava flows

sills and dykes

bedded móberg lapilli tuff and tuff

ankaramite

talus

Dysjarhóll   Pöst   mine

Figure 5.2 A cross section showing the lithostratigraphy at Hvammsmúli.

produced the 10 000-year-old Hamragarðar and Kambagil lava flows (see Fig. 5.1: site 1) and one explosive dacite eruption from the summit crater in 1823.

Locality 5.1 Hvammsmúli

At the base of Hvammsmúli, a small promontory south from Eyjafjalla-jökull, outcrops a 45 m-thick massive pale-grey body of highly olivine–pyroxene phyric ultrabasic rock known as ankaramite, which represents the most primitive rock type of the transitional alkali series. The 0.6 million-year-old Hvammsmúli ankaramite is best exposed at Dysjarhóll, Pöst and in a nearby quarry, and is directly overlain by a 2–4 m-thick unit of bedded tephra and thin lava of the same composition (Fig. 5.2). Originally, the Hvammsmúli ankaramite was interpreted as a sill, but new investigations suggest that it may represent a lava lake formed in association with the overlying ankaramite tephra and lava. The upper part of Hvammsmúli consists of subaerial olivine-phyric basalt-lava flow sequences alternating with thicker units of subglacial hyaloclastite breccias and tuffs.

On the way. The Skógafoss waterfall by the old settlement farmstead Skógar is perhaps the most perfectly shaped waterfall in Iceland, falling off a 60 m-high cliff tucked in between bright green grass-covered slopes.

Locality 5.2 Hofsá River – myth and the truth about the formation of Skógasandur and Sólheimasandur

The Book of Settlement tells the story of a dispute between Þrasi at Skógar and Loðmundur at Sólheimar, two farms on either side of Jökulsá at Sólheimasandur. According to the folktale, one morning Þrasi saw a great flood approaching from the mountains and used sorcery to force it into a more easterly path towards Sólheimar. Loðmundur, who was blind, was told by one of his slaves that an ocean of water was cascading down the mountain slopes and heading their way. Loðmundur knew Þrasi's game, so he used his own powers to redirect the floodwaters westwards in the direction of Skógar. Þrasi responded and Loðmundur repeated his act. This apparently went on for a while until they agreed on a floodpath midway between the farms, where the River Jökulsá on Sólheimasandur now runs directly to the sea. Also, according to one version of the tale, these floods formed the sandur plain of Sólheimasandur.

For a long time it was thought that the floods mentioned in the tale of Þrasi and Loðmundur had formed the whole outwash plain of Skógasandur and Sólheimasandur, because geological evidence appeared to suggest an age within historical times for the main tephra-laden flood deposits on both sandur plains. However, recent findings show that the main flood deposits at Skógasandur are prehistoric (Fig. 5.3) and were formed by a jökulhlaup that accompanied a seventh-century Katla eruption. In this eruption the vents were in the western part of the caldera and within the drainage area of Sólheimajökull outlet glacier. However, younger flood deposits of lesser volume that formed during Settlement age cover part of Sólheimasandur. We also know from tephrochronological studies that the early settlers soon became acutely aware of the awesome nature of Katla, because within 70 years of the arrival of the first settlers it had erupted three times. After being exposed to two relatively small Katla eruptions in the late ninth and early tenth centuries, the newcomers witnessed the largest flood-lava eruption on Earth in the past 2000 years, the 934–40 Eldgjá–Katla Fires. One of these eruptions is likely to have been responsible for the historical jökulhlaups onto Sólheimasandur, which supposedly kindled the dispute between Þrasi and Loðmundur.

Figure 5.3 Schematic cross section of the Hofsá River channel where it dissects the main Skógasandur flood deposits. On the east bank of the river the tephra-rich flood deposits several metres thick and overlay a soil containing many prehistoric tephra layers and a distinct horizon with tree-remains near the top. The ^{14}C age obtained from these tree remains is 1280±100 years (i.e. about AD 600). In the west bank the flood deposits are much thinner (0.5 m) and overlain by soil containing historical tephra layers including the Settlement layer formed around 870. Therefore, the Skógasandur floods occurred well before the first settlers arrived in Iceland.

Locality 5.3 Dyrhólaey and the islands in the sand: volcanoes telling the tale of a different past

Some of the unusual features of the region around Mýrdalur are the mountains Pétursey, Dyrhólaey and Hjörleifshöfði, which stand as isolated crags in a sea of sand and generally a good distance inland from the current coastline. Now on dry land, these structures are remnants of emergent submarine volcanoes formed when these low-lying coastal plains were fully submerged. Since then, the rivers have transported enough debris to raise the area out of the sea and link the former islands with the main landmass of Iceland.

Dyrhólaey (Portland) is the southernmost point of Iceland (63°23'N) and is a heavily eroded submarine volcano of the Surtseyan type (Fig. 5.4). The main part of Dyrhólaey is built of well bedded tuff formed by hydromagmatic explosions and it represents the remains of a larger tuff cone. On the eastern flank of Dyrhólaey, the tuff sequence is capped by compound pahoehoe lava, which in places exhibits cube jointing, indicating water-enhanced cooling of the lava. The lava flows represent the subaerial phase of the eruption when the tuff cone had grown large enough to prevent sea water from accessing the vent. Flowing away from the vent(s), the lava re-entered the sea and cooled rapidly to form the cube-jointed lobes. The Dyrhólaey sequence is typical for Surtseyan eruptions and, if visibility allows, the type-volcano Surtsey can be seen from this location as the southernmost island of the Westman Islands.

Other, perhaps a little more obscure, evidence of submarine volcanic activity is found above Mýrdalur in the móberg mountain above the Skammidalur farm. The mountain is a remnant of a submarine tuff cone made up of bedded lapilli tuff containing isolated blocks of fossiliferous marine sandstones and conglomerates. The fossil-bearing blocks, up to 1 m in diameter, were plucked from the underlying substrate by the Surtseyan explosions in the volcano conduit and incorporated as accidental clasts (i.e. xenoliths) in the tuffs.

Most of the fossils are molluscs, with 15 species of bivalves (lamellibranchia) and 9 species of snails (gastropoda), but they also include species of brachiopods, echinoderms (sea urchin), crabs (crustacea) and worms. Although the Skammidalur fossil assemblage bears a strong resemblance to modern marine fauna around Iceland, almost half are warm-ocean species that now reside farther south in the Atlantic or are extinct. Thus, when these fossil animals were living off the coast of Iceland, the ocean was considerably warmer than it is today. The age of the Skammidalur shell-bearing blocks is 2–3 million years (early Quaternary), as suggested by the similarity of the fauna to that found in the *Serripes* layers at Tjörnes in North Iceland (see p. 140).

| tephra from rootless littoral eruptions | bedded hydroclastic tuff | beach gravel | pillow lava |

Figure 5.4 **(a)** Aerial view of Dyrhólaey (courtesy of Ray Millar, Natural Arch and Bridge Society); **(b)** simplified cross section showing the structure of the Dyrhólaey tuff cone (~100 m deep, ~1500 m wide).

Similar shell-bearing blocks are found sporadically in the móberg tuffs extending from Pétursey in the west to Höfðabrekka in the east, indicating that the volcanic formations of Mýrdalur rest on an extensive sequence of marine sediments that also covers the submerged bedrock shelf to the south, well beyond the Westman Islands. A progressive southward young-ing of the sediments is suggested by the fossil-bearing sandstone xenoliths ejected by the explosive eruptions at Surtsey in 1963–4, which contain fossils of only modern-day species. Thus, some 2–3 million years ago the sea inundated the Mýrdalur area and it is likely that the now-majestic vol-canoes of Mýrdalsjökull and Eyjafjallajökull began their life some 600 000–700 000 years ago in a similar fashion to the Westman Islands, as a cluster of small volcanic islands in the middle of the sea.

Locality 5.4 Mýrdalsjökull volcano, the Katla fissure and Mýrdalssandur
The Mýrdalsjökull central volcano (1450 m) is not only the second largest volcano in Iceland but it has also produced some of the most cataclysmic eruptions in the country's history. Second to the Hofsjökull volcano, with a total volume of 380 km^3, it is capped by Iceland's fourth largest glacier (580 km^2). Mýrdalsjökull volcano has been the locus of activity on the 100 km-long Katla volcanic system over the past 200 000 years and is characterized by basalt magmatism (see Table 1.4). However, it has produced more-evolved magmas in significant quantities, as is evident from scattered occurrences of rhyolite lavas and ignimbrites on the volcano, and dacite tephra layers in the soils of the surrounding lowlands.

The volcano is crowned by a large (~110 km^2) ice-filled summit caldera that hosts the legendary Kötlugjá ("Katla fissure"). According to an old folktale, the name "Kötlugjá" is derived from a cook at the monastery at Þykkvabæjarklaustur, named Katla – a woman true to ancient traditions and the proud owner of magic trousers. The trousers allowed anyone who put them on to run as far and as long as they desired without getting tired. One autumn the shepherd Barði wore the priceless trousers, without permission, when gathering his sheep. When Katla discovered this, she secretly pulled Barði aside and drowned him by the main door in a large barrel of whey. Over the winter, the whey was gradually used up by the residents of the monastery, and Katla realized that eventually her crime would be exposed. As Barði's body was about to emerge, Katla put on her magic trousers, ran onto the Mýrdalsjökull and flung herself into a chasm in the glacier. Shortly thereafter, the chasm erupted, sending a colossal flood (jökulhlaup) in the direction of the monastery. Accordingly, it was believed that Katla had used her witchcraft in an attempt to avenge her fate and, ever since, the chasm has been known as Kötlugjá. Consequently, it has become common practice to refer to the caldera at Mýrdalsjökull central volcano as Katla, and the outbursts that occur within its boundaries as Katla eruptions.

It is not surprising that the activity at Mýrdalsjökull has been linked to the wicked trickery of the supernatural, considering the frequency of Katla eruptions and the hardship and devastation they caused in the surrounding farming communities. With more than 20 eruptions in historical times, Katla is the second most active volcano in Iceland (Fig. 5.5a). It is notorious for its violent eruptions, caused by explosive interaction between the ice and the fire, as the magma tunnels its way through the 400 m-thick glacier. Hence, Katla eruptions are typically hydromagmatic and are accompanied by widespread tephra fall and colossal jökulhlaups consisting of meltwater, ice and volcanic debris. The most recent eruption of Katla was in 1918, which lasted for 24 days (12 October to 4 November).

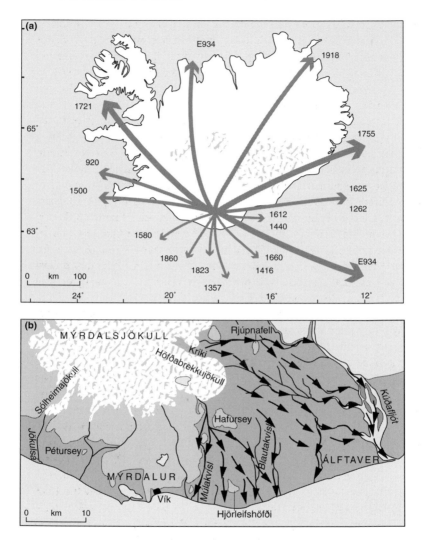

Figure 5.5 **(a)** The tephra dispersal directions for historical Katla eruptions (numbers indicate year of eruption). **(b)** the main flow paths of the jökulhlaup formed by the 1918 Katla eruption.

At about 1 p.m. on 12 October a sharp earthquake was felt in Mýrdalur, and two hours later a 14 km-high eruption column was seen rising from the Katla fissure, dispersing tephra towards the north-northeast in such quantities that it laid to ruin the farm (Búlandssel) in the Skaftártunga district. At about the same time the first jökulhlaup was seen to burst forwards in two places from the outlet glacier Kötlujökull, to the west of Hafursey and at Kriki. This flood covered much of the sandur plain. The jökulhlaup had two main flood paths. At the Hafursey outlet the flood was directed towards the south, following the Múlakvísl and Sandá river beds before cascading onto the sea on either side of Hjörleifshöfði. The Kriki outlet directed the floodwaters eastwards across the plain into the rivers Hólmsá, Skálm and Landbrotsá and then along the course of the Kúðafljót River to the sea (Fig. 5.5b).

A second flood pulse emerged from the Hafursey outlet at 5 p.m., transporting so much ice that it "looked like snow-covered hills were rushing forwards". The floodwaters advanced at velocities of 15–20 km/hour, with a peak discharge of around 200 000 m³/s. The total volume of the 1918 jökulhlaup is estimated at 3–5 km³ and it flooded more than 50 per cent of the Mýrdalssandur, covering 400 km². In the process it dumped 0.5–1 km³ of tephra and other debris onto to the sandur plain, elevating its surface by about 1 m. The 1918 jökulhlaup also added 14 km² of new land to the coastline in front of Hjörleifshöfði. This new land was called Kötlutangi and for some years it was the southernmost point of Iceland. Kötlutangi soon lost its honorary status because the prevailing coastal currents easily eroded the unconsolidated flood deposits.

Katla continued to erupt explosively for the next 23 days, although at decreasing intensity, dispersing tephra to the northwest and southeast of the volcano. The total volume of freshly fallen tephra from the 1918 eruption is estimated at 0.7 km³, roughly equal to 0.3 km³ of magma.

Since the twelfth century the eruptions at Mýrdalsjökull have been centred on the Katla fissure and have progressed in much the same manner as the 1918 eruption described above. These events were all explosive subglacial eruptions producing tephra falls and were accompanied by jökulhlaups onto Mýrdalssandur. However, the magnitude of these eruptions has varied greatly. The smallest, such as those of 1823 and 1860, produced less than 0.02 km³ of tephra and were accompanied by jökulhlaups of small to moderate magnitude. However, the largest, namely the 1262, 1625, 1721 and 1755 eruptions, emitted large quantities of tephra (0.5–1.5 km³), with tephra fall extending to more than 200 km from source and large jökulhlaups of magnitude similar to or larger than the one in 1918. These eruptions were powerful enough to cause tephra fall in Scandinavia and western Europe.

By the same token the magnitude of the damage inflicted by these eruptions on the land and nearby farming communities was significantly greater. For example, over 50 farms in the Fire Districts were abandoned for months and even years in the wake of the tephra fall from the 1755 Katla eruption.

Katla jökulhlaups have caused not only hardship, by destroying farms, cultivated fields and fine pastures in and around Mýrdalssandur, but also significant changes in the environment. In settlement times a coastal lagoon named Kerlingarfjörður, large enough for Viking ships to dock, existed in front of Hjörleifshöfði. The lagoon was destroyed by the first post-settlement jökulhlaup onto Mýrdalssandur, "Höfðárhlaup" in 1179. The lagoon Kúðafjörður in Álftaver, another useful harbour, met a similar fate in the so-called "Sturluhlaup", which occurred as a result of the 1262 Katla eruption. In short, Mýrdalssandur and the neighbouring areas have changed more in the past 1100 years than, for example, the landscape of Western Australia has in 40 million years.

A simple count of Katla tephra layers in prehistoric soils around the volcano reveals that well over 100 basaltic and at least 12 felsic explosive subglacial eruptions have occurred at the Mýrdalsjökull volcano in the past 10000 years. In addition, 5–10 small effusive eruptions have occurred along the northern periphery of Mýrdalsjökull, and the fissure system extending to the northwest has produced two flood-lava eruptions: the 6600 BC Hólmsá Fires and the 934–40 Eldgjá Fires. The latter is a giant among historical lava eruptions on Earth. In total the volcano has produced 30–35 km^3 of tephra and close to 25 km^3 of lava in the past 7000 years, making it the most productive volcano in Iceland.

The largest explosive, and the only Plinian eruption, at Katla occurred in Allerød time (10600 years BC) when more than 10 km^3 of rhyolite rushed out of the volcano, sending pyroclastic flows cascading down the southern flanks and dispersing airborne tephra to more than 1300 km away from the source. The pyroclastic flows are known as the Sólheimar ignimbrite and they outcrop at several localities in Mýrdalur, and the tephra fall deposit forms an important time marker in marine sediments in the North Atlantic, where it is known as ash zone 1. On land in Iceland this tephra fall is known as the Skógar tephra (see p. 157), whereas in other Nordic countries it is referred to as the Vedde tephra, after the lake deposit in which it was first found in Norway. It is probable that the Katla caldera was formed or substantially enlarged by this eruption.

The current repose period in activity of Mýrdalsjökull volcano (i.e. since 1918) is among the longest known in historical times. Seismic unrest does occur from time to time and, as a precautionary measure, the road traffic across the sandur plain is then halted on both sides of the plain. Increased activity within the caldera of Katla volcano began in 1999.

The Fire Districts
Skaftártunga to Öræfi

This excursion more or less follows Highway 1 through the farming districts of Skaftártunga, Síða and Fljótshverfi, then across the Skeiðarársandur outwash plain to Öræfajökull.

Nowhere in Iceland are the awesome qualities of volcanic eruptions more evident than in the Fire Districts. The area is guarded by the Katla volcano in the west, and hidden under ice in the northwest is Grímsvötn, the most active volcano on Iceland. When these volcanoes erupt, they not only spew ash everywhere but they are also accompanied by violent jökulhlaups that temporarily put hundreds of square kilometres of land under water and destroy everything in their path. The region was also the centre stage for the three largest eruptions since the settlement of Iceland: the basaltic fissure eruptions of Eldgjá (934–40) and Laki (1783–4), and the Plinian eruption of the Öræfajökull volcano in 1362. All of these eruptions altered the landscape significantly and had severe consequences for the inhabitants at the time.

Entering the Fire Districts from the west takes you across the desert-like plain of Mýrdalssandur and then through a cluster of low mounds, which are the Álftaver cone group, formed by the rootless eruptions when the western branch of the 934–40 Eldgjá lava flows covered the wetlands around the old Álftaver settlement. At this point, a distinct scarp will appear on the horizon to the northeast, marking a line of an old seacliff dating back to the beginning of the Holocene. At the eastern end, the cliffs are crowned by Mt Lómagnúpur, the highest freestanding cliff face in Iceland, rising to 767 m above sea level.

The flat-lying area in front of the scarp is an old sea floor that over the past 10000 years has gradually emerged from the sea through the combined effort of glacial rivers and basaltic volcanism. Debris dumped by glacial rivers and their jökulhlaups, along with the periodic advance of large basaltic lava flows, have built up the sandur plains. In the process they have added a 110 km-long and 20 km-wide strip to Iceland, almost in a geological instant.

Farthest east is the glacier-covered Öræfajökull volcano, rising to 2119 m at Hvannadalshnjúkur, Iceland's highest peak. Between Öræfajökull and Lómagnúpur, three outlet glaciers extend from the largest icecap in Europe, Vatnajökull, onto the sandur plain and supply meltwater to the many rivers of Skeiðarársandur. The largest valley glacier, Skeiðarárjökull, is also the main route for jökulhlaups from the subglacial caldera lake of the Grímsvötn volcano. As a consequence of the eruption at Gjálp, the most recent jökulhlaup occurred in 1996, when a subglacial fissure erupted through the ice just to the north of Grímsvötn (see pp. 116–117).

The age of the basement rocks is 0.1–4 million years and they belong to the Upper Pleistocene and the Plio-Pleistocene formations. The oldest rocks are exposed at the foothill of Öræfajökull in the east and become progressively younger to the west as they approach the East Volcanic Zone. The basement rocks are predominantly volcanic in origin, but also feature some well exposed glacial and lacustrine sediments. The volcanic rocks principally consist of alternating layers of basaltic lava (grey) and móberg (brown) that are often intercalated with tillites and other types of diamictites. The abundance of móberg tuffs and breccias is highest in the west within the younger sequences, and gradually decreases eastwards. This occurrence simply reflects the increase in the frequency of subglacial eruptions as the influence of the Ice Age glaciers became more dominant. The landscape changes accordingly, from relatively smooth rolling topography with irregular stratification, to steep-sided and regularly stratified mountains separated by deeply incised glacial valleys.

The basement rocks are partly covered by basaltic lava flows produced by Holocene fissure volcanoes and by the extensive sandur plains that are constantly modified and added to by fluctuating glacial rivers. In other areas the basement rocks are covered by wind-lain soil or loess, which is extremely fertile. Nowhere in Iceland is the soil thicker than here, up to 12 m in the Skaftártunga district. The soils contain hundreds of tephra layers and thus provide an excellent record the volcanic history in the region over the past 6000 years. This is well illustrated by the soil profile at Hólmsárbrú (see Fig. 5.1: site 2), a stop well worthwhile for those interested in tephrochronology.

Locality 5.5 Historical flood lava eruptions
The Fire Districts were the scene of the two largest flood-lava eruptions on Earth in the past millennium. These eruptions produced the Eldgjá and Laki lava flows, which together cover 1100 km², an area roughly equal to the size of Greater London (Fig. 5.6). The fact that these two eruptions produced more than half of the magma volume erupted in Iceland over the past 1000 years is a good indicator of their enormous size. These eruptions were truly extraordinary and of catastrophic proportions, as is keenly hinted at by Charles Lyell in is monumental work, *Principles of geology*. Therefore, it is worth considering these eruptions in some detail. Site 3 on Figure 5.1 provides a good vantage point for viewing the vast lava fields of these two eruptions.

Figure 5.6 The volcanic fissures and lava flowfields of the AD 934–40 Eldgjá and 1783–84 Laki eruptions.

The 934–40 Eldgjá eruption The Eldgjá vents form a discontinuous 75 km-long volcanic fissure extending from the Katla volcano in the west, almost to the tip of Vatnajökull in the east (Fig. 5.6). The eruption takes its name from a spectacular 150 m-deep and 8 km-long chasm called Eldgjá ("fire fissure") that occupies the central part of the vent system. However, Eldgjá is an old tectonic graben that was reactivated in the 934–40 eruption. The thick móberg sequence that makes up the walls of Eldgjá is capped by red and black scoria and spatter deposits containing discontinuous lava-like units formed as fountain-fed flows. The scoria and spatter cones at the

109

Table 5.1 Some basic facts on the 934–40 Eldgjá and 1783–84 Laki eruptions.

Eruption	Duration	Episodes	Column height	Volume (km)		Area (lava)	Length (lava)	SO$_2$ output*
				Tephra	Lava			
Eldgjá	3–6 years?	15	>14 km	1.4 km^3	18.1 km^3	781 km^2	70 km	219
Laki	8 months	10	7–13 km	0.4 km^3	14.7 km^3	599 km^2	65 km	122

* Million tonnes.

bottom of Eldgjá are also considered to be primary volcanic features.

The early settlers must have been amazed by the incessant effusion of lava from the Eldgjá fissures, which spread out in two branches that in the end covered about 800 km^2 (Table 5.1). The eruption also featured 15 explosive phases and probably lasted for six years. The explosive activity generated eruption columns more than 14 km high and dispersed tephra over much of southern Iceland farther than 150 km away from the vents. The magnitude of the eruption is best understood by spreading the Eldgjá lavas along a 35 m-wide and 540 km-long strip, equal to the distance from Paris to Berlin; the lava would form a 1000 m-high wall.

The effects of the eruption on the early settlement in Iceland must have been devastating. Scarce records show that the Eldgjá lava flow advanced over vast areas of productive farmland and forced many settlers from their lands, especially within the districts of Álftaver and Síða. Some families were forced to relocate because of thick tephra coverage; others took desperate measures to rake the tephra off their fields. We also know that the lava forced the Skaftá River into a new course closer to the foot of the Síða scarp, just south of its present channel, which follows the margins of the younger Laki lava flow (see Fig. 5.6). Thus, the local environmental effects of the eruption were enormous. The many rootless eruptions, which formed the cone groups of Álftaver and Landbrot more than 40 km away from the actual vents, must have amazed the onlookers, as the lava converted marshy areas and shallow coastal inlets into dry land.

The Eldgjá eruption was the greatest source of volcanic pollution in recent history, exceeding the other two "biggies", the 1783 Laki and 1815 Tambora eruptions, by factor of two. It pumped 219 million tonnes of SO$_2$ into the atmosphere and this gas reacted with atmospheric vapour to produce about 450 million tonnes of H$_2$SO$_4$ aerosols that were spread to remote regions and covered much of the Northern Hemisphere. In comparison, the most climatically significant eruption of the twentieth century, the 1991 eruption at Mt Pinatubo in the Philippines, injected about 10 million tonnes of SO$_2$ into the atmosphere. Scanty information in historical accounts indicates that the eruption had a significant effect on the atmosphere and weather patterns in Europe and the Middle East for several years. However, the intensity of the climatic impact of Eldgjá is not thought to have

110

surpassed that of Laki or Tambora, because the eruption was prolonged and subsequently the sulphur emissions were drawn out over several years.

Good outcrops through the Eldgjá lava are scarce, but are found along the Hólmsá River gorge, at Alviðruhamrar, and in the soil section at Hólmsárbrú, where the lava rests directly on its tephra layer (see Fig. 5.1: site 2). In the thick soil bank along the Skaftá River (see Fig. 5.1: site 4) between the Ytri-Dalbær and Kirkjubæjarklaustur farms, all tephra layers older than Eldgjá are disturbed, having been slanted and even rolled up into folds. However, the younger historical tephra layers in the top of the soil are undisturbed and form sub-horizontal continuous layers. Here the great Eldgjá lava forced its way under the soil cover and, through advance and lava inflation, it disrupted the soil. The historical tephra layers were formed after this event and are therefore undisturbed.

The Landbrotshólar cone group was formed by rootless eruptions when the Eldgjá lava advanced onto the marshy coastal plains in front of the Síða scarp. It is by far the largest cone group in Iceland and at present it covers an area of 60 km². Part of the group is now buried beneath the 1783 Laki lava flow and it may have originally covered 150 km². In 1793 Sveinn Pálsson examined the Landbrotshólar cone group and described it as pyroclastic cones with a lava cap (i.e. welded spatter). He was the first to suggest that it was formed by secondary eruptions.

The 1783–84 Laki eruption   We know more about the Laki flood lava eruption than any other of its kind, because it is described in many contemporary accounts, including several chronicles that contain enough details about the eruption to be regarded as Eldrit ("chronical of an earth fire"). However, none are as spectacular as the Eldrit, "A complete description of the Síða Fires", written in 1788 by the Reverend Jón Steingrímsson at Prestbakki in the Síða district. This chronicle not only contains vivid and accurate descriptions of the events of the eruption on a day-to-day basis and its consequences for the local community, but also demonstrates how intrepid we can be when faced with the extremes of Mother Nature.

In 1783 the people of South Iceland had enjoyed a favourable spring and were looking forward to summer. However, their destiny was about to change. Weak earthquakes in the Skaftártunga district in mid-May were the first sign of what was to come. The intensity of these earthquakes increased steadily and on 1 June they were strong enough to be felt across the region from Mýrdalur and Öræfi. The earthquake activity escalated until 8 June, when a dark volcanic cloud spread over the district, blanketing the ground with ash (Fig. 5.7a). The Great Laki eruption had begun.

Later that day, lava columns more than 1000 m high rose from a new volcanic fissure up in the highlands to the north. The volcanic fumes filled up the atmosphere such that the Sun appeared red as blood and was deprived of its natural shine. The accompanying rainfall contained salty and sulphurous water, causing smarting in the eyes and on the skin.

On 11 June the channel of the mighty Skaftá River dried up and on the following day a large lava stream surged out from the Skaftá River gorge with an incredibly loud cracking noise and rumbles. The volume of this lava was similar to that of the river before it dried up. Recurring tephra falls occurred through June and July, and the lava continued to surge out of the gorge for the next 45 days (Fig. 5.7b). By the time it stopped, the lava had filled up the gorge, which was up to 100 m deep before the eruption, and covered 350 km^2 of the land, including the cultivated areas in front of the gorge and 17 farms.

However, the Laki fissures were not done. On 29 July the heavy tephra fall from the fissures covered the eastern part of the Fire Districts, forcing many farmers off their land (Fig. 5.7a). On 3 August the channel of the Hverfisfljót River dried up and shortly thereafter fresh lava surged out from its gorge (Fig. 5.7b). One can only imagine what went through people's minds when they saw this new flow advancing and threatening to engulf the Síða district completely in lava.

In the autumn the Fire Districts were showered by intermittent tephra falls and the lava continued to flow from the Hverfisfljót River gorge until the end of October, destroying an additional 4 farms and added 250 km^2 to the Laki lava flowfield. The Laki fissures continued to produce lava until 7 February, when the fires at the fissures were finally extinguished.

As if one erupting volcano was not enough, in mid-July 1783 "a sandy and muddy ash" fell in the Fire Districts "from another eldgjá". This was the first of many concurrent subglacial eruptions at the Grímsvötn volcano, which occurred intermittently until May 1785 (Fig. 5.7a).

The Laki eruption (8 June 1783 to 7 February 1784) nearly exceeded its neighbour Eldgjá, forming the second-largest basaltic lava flow in historical times (Table 5.1). The eruption occurred on a 27 km-long fissure delineated by more than 140 vents and craters (see cover illustration and Fig. 1.11c). Ten eruption episodes occurred during the first five months of activity at Laki, each featuring shortlived (2–4 days) explosive eruptions followed by a longer lasting phase of lava emissions. The explosive activity from the eruption produced a tephra layer that covered more than 8000 km^2, the volume of which exceeds that produced by all four Hekla eruptions in the twentieth century. The noxious fumes emitted by the eruption stunted grass growth and killed more than half of the livestock through fluorine poisoning. These consequences ultimately resulted in the disastrous "Haze Famine" that killed 20 per cent of the population.

Figure 5.7 (a) Main dispersal directions of plumes and tephra falls from the Laki fissures during the summer and fall 1783. Also shown are timing of eruptions at the Grímsvötn volcano; (b) growth of the Laki lava flowfield reconstructed using information from the contemporary accounts on the position of the lava flow front at the given dates.

Figure 5.8 Simplified vertical section from Iceland and across mainland Europe showing the eastward transport of the sulphuric aerosol plumes from Laki.

The effects of the Laki eruption extended well beyond Iceland. Its columns pumped 100 million tonnes of SO_2 into the westerly jet stream, which governs the atmospheric circulation above Iceland, producing sulphur-rich plumes that were dispersed eastwards over the Eurasian continent and north into the Arctic. The sulphur dioxide reacted with atmospheric vapour to produce 200 million tonnes of sulphuric acid (H_2SO_4) aerosols, of which about 175 million tonnes were removed in the summer and autumn of 1783 by subsiding air masses within high-pressure systems (Fig. 5.8). This removal formed the infamous dry fog that hung over the Northern Hemisphere for more than five months. The acidity of the dry fog was such that it caused considerable damage to vegetation and crops all over Europe and stunted tree-growth in Scandinavia and Alaska. Furthermore, the following winter is one of the most severe on record in both Europe and North America. The eighteenth-century temperature records from these regions suggest that the annual cooling that followed the Laki eruption was about −1.3°C and lasted for two to three years.

Locality 5.6 The hyaloclastite flows of the Quaternary Síða Group
The Plio-Pleistocene strata of the Síða and Fljótshverfi districts that form the old seacliffs above Kirkjubæjarklaustur consist of a 700 m-thick volcanic succession called the Síða Group. It is dominated by 14 large-volume (up to 30 km³) subaqueous hyaloclastite lava flows emplaced onto a submarine shelf. The Síða Group unconformably overlies an eroded sequence of flat-lying lavas and sedimentary rocks (the Fljótshverfi Group), formed before the Ice Age set in. The distribution of individual flows appears to have been mainly controlled by the existing topography, as is illustrated by the units that make up Mt Rauðaberg in Fljótshverfi.

114

The lower part of each hyaloclastite flow consists of columnar jointed lava (5–30 m thick), grading laterally and vertically into pillow lava or cube-jointed lava (5–40 m thick), which in turn is overlain by kubbaberg breccia 10–120 m thick. Each flow sequence is typically capped by a 15 m-thick unit of well bedded lapilli tuff. The lava at the base of these flows often exhibits spectacular columnar joints, such as those forming the pavement-like Kirkjugólf ("church floor") at Kirkjubæjarklaustur and the readily accessible cliffs of Dverghamrar below the Foss farm. The cube-jointed lava typically features poorly developed columns with irregular orientation, although locally these fan out to form regular rosette-jointed bodies. The hyaloclastite flows are separated by 5–50 m-thick sedimentary units consisting of debris-flow deposits interbedded with marine sandstones and mudstones.

In the past 10 000 years the rivers have cut many gorges and gullies into the Síða Highlands, but none as spectacular as the Fjaðrárgljúfur gorge. Here the rather innocent-looking Fjaðrá River has carved out a 100 m-deep ravine into the hyaloclastite flows and other formations of the Síða. In early postglacial times a sizeable lake occupied the valley above Fjaðrár-gljúfur (the flat-top benches along the valley side are the strandlines formed by that lake) and was kept at bay by an erosion-resistant threshold. At the time, the gorge did not exist and a much larger river flowed down the gently sloping hills before cascading off the cliffs of the Síða scarp. By shear power of flow the river rapidly carved out the gorge back into the mountains. Eventually it breached the threshold at the mouth of the valley and in so doing drained the lake, changing the conditions to those of today.

Mt Lómagnúpur (767 m) features the highest cliff face in Iceland, with a vertical rise of more than 600 m. It forms the southern end of the Björninn mountain complex, which is built of Pleistocene móberg and lavas. The cliff itself was carved out by glaciers and later modified by coastal erosion. At the base of Lómagnúpur in the west are deposits from two rock avalanches, one of unknown prehistoric age and another that was formed by an avalanche in July 1789; the latter occurs as a hummocky terrain just inside the southern tip of the mountain. Highway 1 traverses the outermost lobes. Ebenezer Henderson examined this rock avalanche in travels around Iceland in 1814–15 and gave the following description:

> Turning round the extremity of the mountain [Mt Lómagnúpur], which hangs almost directly overhead, at the angle of two ranges of ancient buildings we fell in with many heaps of stones and immensely huge masses of tuffa [móberg], which have been severed from the mountain, and hurled down into the plain, during the rockings occasioned by an earthquake in 1789.

115

It is known from other sources that this rock avalanche occurred in the early hours of the day, sometime in July 1789, and the earthquake mentioned by Henderson was probably induced by the avalanche rather than being its cause. The scar left behind by the avalanche is clearly visible near the clifftop, but on the ground the avalanche deposit forms a 650 m-wide fan. Approximately 2.5 million m^3 (or about 6 million tonnes) of material were displaced in the event.

Locality 5.7 Skeiðarársandur – volcanism in Vatnajökull and the glacial outwash plains

Skeiðarársandur (1000 km^2) is the largest sandur plain in Iceland. It is the alluvial fan of Skeiðarárjökull, the largest outlet glacier on the south side of Vatnajökull icecap, and has an ablation area of 1200 km^2. Skeiðarársandur features a complex braided river system that can broadly be divided into two main systems: the Núpsvötn–Súla–Gýgjukvísl system, which emerges from the western side of Skeiðarárjökull, and the Skeiðará system, which emerges from the eastern side. Although these river systems relentlessly carry large quantities of debris onto the sandur plain, their contribution to its construction is subsidiary when compared to the transport of material by the catastrophic jökulhlaups that rush the sandur plain at regular intervals. The jökulhlaups at Skeiðarársandur originate in two ways: by sudden emptying of the ice-dammed marginal Lake Grænalón and by draining of the caldera lake at Grímsvötn.

Grænalón is an ice-dammed marginal lake that occupies an ice-free tributary valley on the west side of Skeiðarárjökull (see Fig. 5.7a). The lake is fed by surface runoff from the surrounding mountains. When the water in Grænalón reaches a critical level (i.e. 90% of the glacial thickness) it has enough momentum to overcome the ice barrier. The water penetrates under the ice, and subglacial tunnels are carved out by the frictional heat generated by the flowing water. This results in abrupt draining of the lake and sudden floods in the rivers Súla and Núpsvötn. Ice-dammed lakes such as Grænalón are usually drained at fairly regular intervals, typically every 1–2 years.

In chronicles from the sixteenth and seventeenth centuries, Grímsvötn ("lakes of Grímnir") are said to be lakes located within the glacier Gríms-vatnajökull and possess the peculiar nature of spelling "Earth fires" (i.e. volcanic eruptions) that bring about large jökulhlaups onto Skeiðarár-sandur. Grímsvatnajökull is believed to be the original name of the largest glacier in Europe, which we now know as Vatnajökull. In the eighteenth and nineteenth centuries the knowledge of the position of these erupting lakes was uncertain. It was not until they were rediscovered in the early twentieth century that their exact location was firmly established.

Grímsvötn is a subglacial lake that occupies the summit caldera of the most active central volcano in Iceland, which bears the same name. It is located at the head of the outlet glacier Skeiðarárjökull (see Fig. 5.7a). Grímsvötn volcano features one of the most powerful geothermal systems in Iceland, with an average output of 4000–5000 megawatts. The geothermal heat melts the ice, and the meltwater accumulates in a 600 m-deep caldera, forming a lake that is sealed by a floating ice shelf. The ice melting gradually increases the volume of the lake, elevating the water table and the surface of the floating ice shelf. When the surface of the lake reaches a critical level, the water pressure is sufficient to open a 50 km-long subglacial waterway extending from Grímsvötn to the edge of Skeiðarárjökull. When this happens, up to 80 per cent of the lake is drained in a jökulhlaup that typically lasts for one to three weeks. Normally the Grímsvötn caldera lake is drained every five to ten years and the jökulhlaups burst onto the sandur plain through the outlet of Skeiðará.

The total volume of the water expelled by these jökulhlaups is in the order of 1–3 km³, with a maximum discharge of 5000–10 000 m³/s. However, every so often this "normal" flood pattern is interrupted by volcanically induced jökulhlaups, such as the one generated by the subglacial eruption at Gjálp in 1996. Eruptions such as Gjálp melt large quantities of ice quite quickly and the meltwater is discharged directly into the Grímsvötn caldera. This results in a rapid rise of the water level in the caldera lake and in catastrophic jökulhlaups that burst through the outlets of Skeiðará, Súla and Gýgjukvísl. These jökulhlaups typically cover the whole of Skeiðarársandur and dump large volumes of debris and gigantic blocks of glacier ice onto the sandur plain (Fig. 5.9).

Locality 5.8 Skaftafell national park and Öræfajökull volcano
East of Skeiðarársandur is the Öræfajökull stratovolcano, one of the three giants among Icelandic volcanoes. Towering over the Skaftafell national park, Öræfajökull rises with an average slope of 15° to a summit plateau with an average height of 1850 m. The plateau is an ice-filled summit caldera with an area of 14 km². Several peaks rise above the caldera, including the rhyolite dome Hvannadalshnjúkur, Iceland's highest peak (2119 m). It is the tallest volcano in Iceland and the third largest in terms of volume (285 km³).

Öræfajökull is the most magnificent glacier area in Iceland. From the ice-capped summit, outlet glaciers plunge down the outer slopes as ragged icefalls that have carved steep-sided valleys into the cone. Many of the outlet glaciers extend to the lowland plains at the foot of the mountain, where they have formed spectacular moraines, such as the end moraines, 100 m-high, in front of Kvíárjökull. Many of the outlet glaciers feature

Figure 5.9 The 1996 jökulhlaup on Skeiðarársandur.

remarkably regular **ogives** that have earned them a place in the geological history of Iceland. On 11 August 1794, when the naturalist Sveinn Pálsson saw the ogives on Hrútárjökull from the caldera rim, he realized that glaciers had the fluid properties of highly viscous matter and that the outlet glaciers moved down slope by flow.

Most of the rock formations that make up Öræfajökull are younger than 0.7 million years and are typical of the Upper Pleistocene Formation. The volcanic succession is principally made of pillow lavas and hydroclastic tuffs and breccias, along with basalt and andesite lava flows. Rhyolites are present in significant amounts, as both intrusive and extrusive units. Several of the peaks rising from the caldera rims are remnants of rhyolite lava domes. These volcanic units are intercalated with lesser volumes of tillites, glaciofluvial and glaciolacustrine sediments.

The most spectacular sediment formation within the Öræfajökull succession is the fossiliferous Svínafell interglacial sedimentary beds that outcrop at the base of Mt Svínafellsfjall (Fig. 5.10). The sediment sequence is 120 m thick and consists of well bedded fine-grain lacustrine sandstone and siltstone, which contain pollen grains and leaf imprints of alder, birch, rowan, blueberry and crowberry heather, grasses and ferns. The sediments rest unconformably on a 100 m-thick series of basalt-lava flows of Lower Pleistocene age (>0.7 million years) and are overlain by a thick

Figure 5.10 Lithostratigraphy of Mt Svínafjall, showing the position of the fossiliferous lacustrine sediments.

tillite bed and a series of pillow lavas, subaerial basalt-lava flows and hyaloclastites. The age of the Svínafell sediments is probably about 0.5 million years and the fossilized assemblage indicates vegetation similar to the present-day fauna in Iceland.

Tephrochronology shows that the postglacial volcanic activity at Öræfajökull has been almost exclusively explosive and has not added much to the height and volume of the volcano. Three postglacial flank eruptions produced small scoria cone rows and two lava flows on the southern slopes. The first historical eruption at Öræfajökull occurred in early June 1362 and was described in a contemporary chronicle as follows:

> Volcanic eruptions in three places in the South and kept burning from flitting days [i.e. early June] until the autumn with such monstrous fury as to lay waste the whole of Litla-Hérað, as well as a great deal of Hornafjörður and Lónshverfi districts, causing desolation for a distance of some hundred miles. At the same time there was a glacier burst from Knappafellsjökull into the sea, carrying such quantities of rocks, gravel and mud as to form a sandur plain where there had previously been 30 fathoms [50 m] of water. Two parishes, those of Hof and Rauðilækur, were entirely wiped out. On even ground, one sank in the sand up to the middle of the legs, and winds swept it into such drifts that buildings were almost obliterated. Ash was carried over the northern country to such a degree that footprints became visible on it. In addition to this, pumice was seen floating off the west coast in such masses that ships could hardly make their way through it.

Information from other contemporary annals is similar, one of them stating that the floods accompanying the eruption swept away all buildings within the rectory of Rauðilækur except the church. Yet another states that "no living creature survived except one old woman and a mare".

The scanty information from the contemporary chronicles is supplemented by volcanological studies of the eruption products, which show that the vents of the 1362 eruption were within the summit caldera and that the main phase of the eruption was purely explosive. It began with a powerful Plinian event producing 40 km-high eruption columns dispersing 10 km^3 of rhyolitic tephra towards the east-southeast (Fig. 5.11). The regular and unidirectional distribution of the tephra fall indicates that the initial phase was brief, but powerful enough to cause tephra fall on mainland Europe and in Scandinavia. The 1362 Öræfajökull is the largest explosive rhyolite eruption in Iceland in historical times and its magnitude was similar to that of the 1991 eruption of Mt Pinatubo in the Philippines.

Figure 5.11 Dispersal of the tephra fall deposit from the 1362 Öræfajökull eruption.

The 1362 eruption was accompanied by catastrophic floods, which emerged from underneath the outlet glaciers Falljökull and Rótarfjalls-jökull on the west flank of Öræfajökull and destroyed a few farms in the vicinity of the mountain. However, the main devastation was caused by tephra fall. The prosperous rural settlement, Litla-Hérað, which consisted of at least 30 farms, was laid waste and remained abandoned for decades. When a revival came at last, the district assumed a new name. It was called Öræfi, the name it still carries today and which literally means "waste-land". The second and most recent historical eruption of Öræfajökull occurred on 3 August 1727 and lasted until April or May 1728.

Locality 5.9 Breiðamerkursandur and Jökulsárlón glacial lagoon
Breiðamerkursandur is the collective name for the outwash plains in front of Hrútárjökull, Fjallsárjökull and Breiðamerkurjökull. These glaciers reached their maximum postglacial extent during the latter half of the nineteenth century, forming a classical piedmont glacier. In 1894 the shortest distance between edge of Breiðamerkurjökull and the coastline was only 256 m. At the River Jökulsá, the distance between the glacier and the sea was about 1 km. Since the mid-1890s these glaciers have been retreating. Between 1894 and 1968 the front of the piedmont glacier retreated about 2 km, breaking it up into the three outlet glaciers we see today. At the time of settlement in Iceland (874–950), the front of Breiðamerkurjökull was at least 10 km behind the 1894 end moraines. At that time, two farms were established on the western side of Breiðamerkursandur. One of them, named Fjall, was located at the foot of Breiðamerkurfjall, and the other, Breiðá (with a church), was located a little farther to the east. Both farms were destroyed between 1695 and 1720 by the advance of the glaciers during the cold snap of the Little Ice Age. The retreat of the Breiðamerkurjökull during the twentieth century since about 1930 has resulted in the formation of the glacial lagoon, Jökulsárlón, which reaches a depth of about 110 m below sea level.

Chapter 6

The southeast and east

General overview

There is a striking change of scenery to the east of Breiðamerkurjökull as we pass into the older Tertiary Basalt Formation (Fig. 6.1a). Here the main outlet glaciers of the Quaternary icecap have honed and enlarged pre-existing fluvial ravines to carve deep U-shape valleys into the gently dipping basalt-lava plateau, thereby producing a classical alpine landscape. Many of these glacially carved valleys are so deep that they reach well below sea level and form a fjord landscape, which is the most distinctive feature of East Iceland. Typically, the main outlet glaciers were fed by many smaller tributary glaciers. These tributary glaciers carved out the hanging valleys and **corries** (cirques) that line the mountain slopes above the valleys and fjords. Good examples are found at Stöðvarfjörður and Fáskrúðsfjörður, where in some instances the tributary glaciers have cut so far back into the mountainside that the corries of these adjacent fjords are separated by only an extremely thin serrated ridge known as an arête.

Eastern Iceland provides a continuous 8500 m-thick succession through the Tertiary Basalt Formation. This succession covers about 10 million years of geological history (from about 13.5 to 3.5 million years ago), indicating an average accumulation rate of 850 m per million years. The Tertiary Basalt Formation was constructed in much the same way as the volcanic successions in the active volcanic zones and it comprises the same geological elements, with the exception of subglacial volcanic and sedimentary rocks. The volcanic system is the principal building block of the succession and it occurs as 10–30 km-wide and 50–100 km-long lenticular stratigraphical units. So far, a total of 17 extinct volcanic systems have been mapped, each made up of an extinct central volcano and a dyke swarm, which in the Tertiary Formation are exposed in vertical section as a swarm of dykes (Fig. 6.1a). The successions that make up the central volcanoes comprise abundant (>20%) intermediate and rhyolite volcanic rocks and therefore are easily distinguished from the sequences dominated by basalt lava, which are produced by the associated fissure swarms.

á house, æ eye, au feuille, ð that, ei day, é yet, hv quick, í seen, ö turn, ú soon, þ think

Figure 6.1   (a) The main geological features of East Iceland. Also shown are excursion routes, stops and sites. **(b, overleaf).**

Figure 6.1 (b) Occurrence and distribution of extinct volcanic systems in East Iceland. The central volcanoes are named.

We present two excursions in Southeast and East Iceland. The first one takes us across Southeast Iceland from Breiðamerkurjökull to Álftafjörður. The second excursion deals with the Tertiary Basalt Formation in East Iceland from Berufjörður to Fljótdalshérað.

Suðursveit–Lón–Álftafjörður

This excursion follows Highway 1 from Breiðamerkurjökull to Álftafjörður and includes the districts of Suðursveit, Hornafjörður and Lón.

Locality 6.1 Suðursveit – valley glaciers and alpine landscape

At Suðursveit the cultivated lowlands rise only a few metres above sea level. To the south, they are separated from the ocean by a series of coastal lagoons and to the north the steep and rugged mountains shelter them from the Vatnajökull icecap. Despite their lush green, the narrow stretches of farmland were originally sandur plains, built from debris transported by glacial rivers such as Steinavötn, Kolgríma and Hornarfjarðarfljót.

The area is a classic example of alpine landscape, where the mountains are incised by a series of steep-sided valleys carved out by descending outlet or valley glaciers. The glaciers have disappeared from some valleys, whereas others are filled to the brim by outlet glaciers from the Vatnajökull icecap (Fig. 6.1a). The 10 km-long Kálfafellsdalur is a spectacular example of these valleys, where the Miðvötn–Steinavötn River runs down the narrow sandur plain flanked by 1000–1500 m-high mountains, which are among the tallest in Iceland. At the bottom of the valley, a small glacier, Brókárjökull, cascades off a 800 m cliff face to the valley floor, forming a stunning icefall. Occasionally, large ice blocks tumble down the steep face of the glacier, accompanied by a rumble that reverberates throughout the valley. The surrounding mountains are composed almost entirely of basalt flows, which originally formed an extensive lava plateau that later was carved out by the Ice Age glaciers. These flows are some of the youngest lavas of the Tertiary Basalt Formation and they formed by repeated effusive eruptions about 5–6 million years ago. Deep in the valley many inclined dykes cut through a series of rhyolite and basalt lavas, indicating the presence of an extinct central volcano.

Farther east, the large valley glaciers Skálafellsjökull, Heinabergsjökull and Fláajökull stretch their snouts onto the coastal sandur plain and in doing so demonstrate what the situation was like at Kálfafellsdalur at the time of its formation.

On the way. Mt Borgarhöfn (or Hestgerðismúli), the promontory to the east of Kálfafellsdalur, is crowned by a magnificent colonnade. These are the columnar joints of a 15–20 m-thick basalt-lava flow that is part of the so-called Múli Formation. This formation belongs to the Plio-Pleistocene Formation (<3 million years old) and comprises the youngest rocks in this region (Fig. 6.1a). The Múli Formation rests on an eroded Tertiary basement and it consists of the above-mentioned columnar lava, which grades downwards into pillow lava and kubbaberg breccia. These volcanic rocks are capped by a thin tillite bed and are underlain by well bedded lake sediments. Below the lake sediments is a thin diamictite bed and then a unit of layered móberg tuff. At the base is a thick tillite unit that rests directly on the eroded Tertiary basement. The Múli Formation was formed in a valley eroded into the Tertiary basement, and the two tillite beds at the foot and top of the sequence indicate that on at least two occasions the valley was congested by glaciers. The sequence also shows that at least two volcanic eruptions contributed to the construction of the Múli Formation. The lower one, the móberg tuff, is clearly a product of a subglacial eruption that occurred when the first glacier occupied the valley. The latter is a lava flow that rests directly on lake sediments, indicating that the lava flowed into a small lake located in front of a valley glacier. The pillow lava and kubbaberg breccia at the base of lava were formed when the lava initially came into contact with the lake water, and their thickness shows that the water depth was not more than 2 m. When this pedestal had been formed, the rest of the lava advanced forwards on dry land and, as the lava cooled and solidified, formed the spectacular columnar joints. Finally, the glacier rushed forwards once again, forming the topmost tillite bed.

Locality 6.2 Hornafjörður to Lón – through a volcano
Traversing the region from Hornafjörður across Lón to Álftafjörður is probably as close as one can get to travelling through a volcano. It is difficult to imagine that the peaceful countryside of Lón was once the boiling hot roots of three volcanoes, now extinct, located at depths of 2–3 km below the surface.

Here, erosion has removed much of the extrusive rock and exposed deeper intrusions. The larger gabbro intrusions, such as those exposed at Viðborðsfjall, Vesturhorn and Austurhorn (Fig. 6.1a), represent ancient basaltic magma chambers that provided the fuel to some of the Tertiary central volcanoes in the region when they were active about 6–8 million years ago (Fig. 6.2). Eruptions from these magma chambers formed basaltic lavas. Several granophyre intrusions (e.g. at Ketillaugarfjall, Slaufrudalur, Vesturhorn and Austurhorn) are also exposed here. Typically, they are situated a little higher in the stratigraphy and above the gabbro intrusions. These granophyre intrusions represent smaller holding chambers

Figure 6.2 Simplified cross section of a Tertiary central volcano and its roots, showing the relative position of basaltic (gabbro) and rhyolite (granophyre) magma-holding chambers and distribution of dyke intrusions.

that at the time contained rhyolite magma formed by partial melting of the surrounding basaltic crust. Eruptions from these chambers were responsible for formation of the rhyolite lavas and pyroclastic flows.

The Austurhorn intrusion is a composite gabbro and granophyre intrusion that outcrops over 11 km² and represents one of the solidified magma chambers in the Lón region (Fig. 6.1b). The eastern part of this intrusion consists of a net-veined complex, formed by intimate mixing of mafic and acid magma (Fig. 6.3). In the net-veined complex the mafic magma occurs as pillow-like or angular masses enclosed in and veined by acid magma (granophyre). This complex was formed as the hotter mafic magma intruded into semi-molten granophyre magma situated near the chamber roof. The fine-grain margins of the pillow-like bodies show that the mafic magma was quenched against the colder granophyre magma.

Figure 6.3 Geology of the Austurhorn gabbro intrusion: rock types and their outcrop pattern.

127

On the way. Travelling across Lónsheiði we enter the domain of the Álftafjörður volcano (Fig. 6.1b: site 1), which features the spectacular Mælifell caldera, a subsidiary structure on the southern margin of the main caldera of the Álftafjörður volcano. The Mælifell caldera is 2 km in diameter and is made up of inward-dipping welded tuffs, agglomerates and tuffaceous sediments, along with rhyolite and andesite lava flows. The welded tuffs occur as thin sheets of pitchstone (obsidian-like rock) and as much thicker units of rhyolite lava-like masses. The pitchstone sheets contain basalt-glass (melt) inclusions, indicating concurrent eruption of rhyolite and basalt magmas. These relationships, along with those observed at Austurhorn (see above), imply that the rocks at Mælifell were formed in eruptions triggered by intrusion of hot mafic melt into a cooler felsic magma body.

The eastern fjords

This excursion takes us across the Tertiary Basalt Formation in East Iceland from Berufjörður to Fljótdalshérað.

Locality 6.3 Breiðdalur central volcano
The Breiðdalur volcano is a classic example of an extinct and eroded volcanic centre. It has a volume of 400 km^3 of mafic, intermediate and felsic lavas and pyroclastics that have a maximum thickness of about 2000 m. The central vent area of the Breiðdalur volcano is manifested as a profusion of rhyolite and andesite lavas and pyroclastic rocks, which are flanked by thin mafic lava flows dipping (5–10°) away from the central vent area (Fig. 6.4). At the time of activity, the Breiðdalur volcano was a 500–1000 m-high cone rising above the surrounding lava plain, which was constructed by eruptions of the associated fissure swarm. The Röndólfur Group is a series of conspicuous peaks (i.e. Stöng, Slöttur, Röndólfur, Smátindur and Flögutindur) on the eastern flank of the volcano, made up of rhyolite lavas and feeder dykes.

Locality 6.4 The Tertiary dyke swarms
The Tertiary succession is dissected by many subparallel basalt dykes, such as those cutting through the cliffs above the Karlsstaðir farm on the north side of Berufjörður (Fig. 6.1a: site 2). Here the near-vertical dykes are 3–6 m thick and they display the horizontal jointing arrangement that is a characteristic feature of dyke intrusions. The joints form because the magma shrinks upon cooling and solidification; they are always perpendicular to the cooling surface (Fig. 6.5a). When the magma intrudes the

Figure 6.4 Geology of the Breiðdalur volcano.

bedrock as a vertical dyke, the joints have horizontal orientation because the principal cooling surfaces are vertical.

The dykes in the Tertiary Basalt Formation typically occur in swarms 5–15 km wide and over 50 km long, and represent the roots of once-active fissure swarms like those seen on the surface within active rift zones. The dyke swarms are typically associated with central volcanoes, such as the Breiðdalur volcano described above (see Fig. 6.1b). Each swarm consists of

Figure 6.5 **(a)** Dykes (arrows) cutting through the basalt lava succession in Fáskrúðsfjörður, East Iceland. **(b)** Map showing variation in intrusion densities across Reyðarfjörður dyke swarm, East Iceland.

hundreds of dykes. The number density is highest along the central axis of the swarm and decreases laterally to the margins (Fig. 6.5b). It also increases with depth below the original surface, because only about 25 per cent of the dykes reach the surface and become feeders to eruptions; the remainder are arrested within the crust, where they cool and solidify.

By measuring the collective thickness of dykes along a horizontal line over a given distance (e.g. 1000 m), one can calculate how volume is occupied by the dykes. Such measurements in East Iceland show that the dykes most commonly make up 5–6 per cent, meaning that the cumulative thickness of the dykes amounts on average to 50–60 m per 1000 m of crust. In the centre of the swarms, they can occupy up to 28 per cent of the crust. The significance of these measurements is revealed when we think of the plate movements. The space occupied by the dykes is a direct measure of how much spreading occurred because of crustal rifting when these systems were active.

Locality 6.5 The Sandfell laccolith
Along the south side of Fáskrúðsfjörður, Mt Sandfell attracts immediate attention because of its pale rocks, in stark contrast with the dark basalt lavas that make up the bulk of the stratigraphical succession here. The pale rock at Sandfell is rhyolite. The steeply dipping strata adjacent to the rhyolite body show that it intruded the basalt-lava series; it is a stunning example of shallow intrusions known as laccoliths (Fig. 6.6). At the time, the slowly rising rhyolite magma came within 500 m of the surface before it began to accumulate as an intrusion. The magma made space for itself by lifting up the overlying strata, forming a dome-shape hill that in the end rose 400–500 m above its surroundings. A small amount of magma made it to the surface through cracks in the up-domed roof. This type of intrusion is also known as a cryptodome, because of its resemblance to rhyolite lava domes, which, however, are formed by extrusions onto the surface.

Locality 6.6 The Skessa tuff and other Tertiary ignimbrites
The Skessa rhyolite tuff is the largest known ignimbrite in Iceland (Fig. 6.7a). It can be found in outcrop over an area of 430 km^2 and has a volume of 4 km^3, which exceeds the volume of the rhyolite tephra fall (3 km^3) produced by the Plinian eruption at Hekla 2900 years ago. It was formed by a pyroclastic flow that originated from an explosive Plinian eruption at vents located below Mt Röndólfur in the Breiðdalur region. Indeed, it is possible that this eruption initiated the growth of the spectacular Breiðdalur central volcano (see p. 129). The Skessa tuff is thickest closest to its source vents (about 10–12 m) and gradually tapers off to less than 2 m at its distal end just north of Reyðarfjörður. Thus, the pyroclastic flow swept

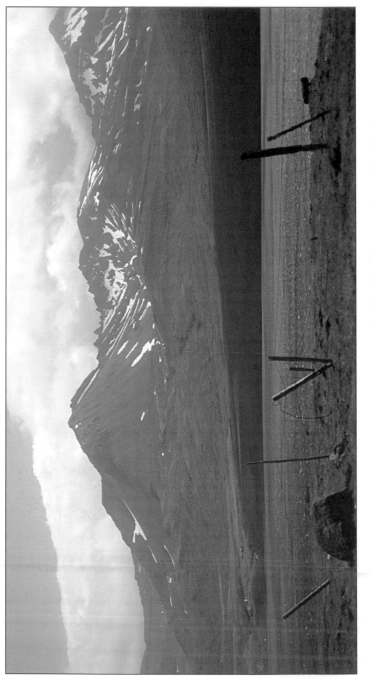

Figure 6.6 Sandfell laccolith. Steeply dipping strata on either side of Sandfell were pushed up by the intrusion.

Figure 6.7 The Skessa tuff.
(a) A map showing the dispersal of the ignimbrite sheet.
(b) The internal structure of the ignimbrite. At the base is a veneer of non-welded tuff that is overlain thin pale-grey glassy welded tuff. The grey welded tuff grades sharply upwards into thicker pinkish stony welded tuff, which is capped by a thin layer of non-welded tuff.
(c, overleaf)

northwards across the lava plains and the hot murky cloud overwhelmed about 500 km² of country.

At its type locality, about 500 m up the north face of Mt Skessa in Reyðarfjörður (Fig. 6.7a), it is the only felsic rock exposed in a 1000 m-thick section of basalt lavas. Here the Skessa tuff occurs as a greenish layer about 3 m thick, sandwiched between two basalt-lava flows. However, here and in outcrops closer to the source vents, the ignimbrite exhibits a distinctive internal layering, as shown in Figure 6.7b, whereas farther to the north this layering is absent.

133

(c)

eruption column

pyroclastic flow

CRUST

RHYOLITE MAGMA

mobilized

rhyolite feeder dyke

basalt rising into rhyolite

basaltic magma

net-veined complex

Figure 6.7
(c) The intrusion of hot mafic melt into cooler rhyolite magma acts as a trigger to ignimbrite-forming Plinian eruptions.

This internal layering results from welding of parts of the ignimbrite during and immediately after its emplacement. In other words, upon deposition, the interior of the ignimbrite was still hot enough for the tephra fragments to be fused together and cause the highly vesicular pumices to collapse and flatten and thus form the flame-like **fiammés** that give the rock a distinct flow-like appearance. This process also hardens the rock and gives the welded part a lava-like appearance, but the **eutaxitic texture** and presence of fiammé reveal its true origin. In general, the thickness of the welded horizon increases to the south towards the source vents. It also shows that, when the ignimbrite was emplaced at Skessa some 30 km north of the source vents, it was still very hot. The general distribution of different rock types in the Skessa tuff with distance from source is depicted in Figure 6.7b.

In some places (e.g. Tinnudalur, Fig. 6.7a), tree moulds are preserved in the non-welded base of the ignimbrite. This shows that the pyroclastic flow decimated everything in its path, and this eruption must have had a devastating impact on the vegetation at the time.

About 2 per cent of the fragments in the Skessa tuff are vesicular basalt clasts, showing that a little mafic magma was erupted with the rhyolite. This suggests that sudden heating of rhyolite magma at depth by much hotter mafic magma triggered the Plinian eruption that produced the

Skessa ignimbrite (Fig. 6.7c). Ignimbrite sheets are fairly common in the Tertiary Basalt Formation; they are very similar in their make-up to the Skessa tuff and probably formed by a similar mechanism. In East Iceland the rhyolitic rocks make up 10–12 per cent of the stratigraphical succession, and the ignimbrites represent about a third of that volume.

Locality 6.7 Lögurinn
In the district of Fljótdalshérað, Lögurinn ("the loch") is East Iceland's largest lake and the third largest in Iceland as a whole. It is about 25 km long, 53 km^2 in area, with a maximum depth of 112 m. The lake is in a U-shape glacial valley that extends more than 90 m below present-day sea level. Tales from the Middle Ages tell a story of a monster that supposedly lives in Lögurinn, similar to the one in Loch Ness in Scotland. Modern-day glimpses are rare and the sighting of the monster at the surface appears to be directly related to events when methane gas escapes from the bottom sediments and bubbles up through the water column.

Chapter 7

The northeast

General overview

The main feature of Northeast Iceland is the North Volcanic Zone and the Tjörnes Fracture Zone that links it with the offshore Kolbeinsey Ridge (Fig. 7.1). There are five active volcanic systems within the 70 km-wide and 200 km-long North Volcanic Zone, including the infamous Krafla and Askja systems (see Fig. 1.5). The Tjörnes Fracture Zone is a 70 km-wide and 120 km-long belt of faults that connects the North Volcanic Zone with the southern tip of the submarine Kolbeinsey Ridge (see Fig. 1.2). It is an active fracture zone, and faulting associated with displacement of the crust is capable of producing earthquakes of magnitude 6–7 on the Richter scale.

We present two excursions in Northeast Iceland. The first visits the canyon country of Jökulsá á Fjöllum and the remarkable sediment succession at Tjörnes. The second excursion is centred on the Lake Mývatn area – one of the most impressive natural wonders in Iceland.

Jökulsá–Tjörnes

This excursion begins at the junction of Highway 1 and Road 864, and follows the canyon of Jökulsá á Fjöllum north to Ásbyrgi. Then it takes us along Road 85 across Axarfjörður to the Tjörnes Peninsula and the town of Húsavík.

Locality 7.1 Jökulsárgljúfur, Dettifoss and Ásbyrgi

Jökulsárgljúfur is the canyon of the River Jökulsá á Fjöllum and it stretches for 30 km from the Selfoss waterfall to the place where the river emerges from the gorge at the head of the sandur plain at the bottom of Öxarfjörður. It offers some of the most magnificent scenery in Iceland. Lavish vegetation thrives in sheltered spots among the regular jointed basalt-lava flows that make up the canyon walls, which sometimes resemble a giant stone-pillar fence erected by forces of another world. The crystal-clear spring

á house, æ **eye**, au feuille, ð **that**, ei day, é yet, hv **quick**, í seen, ö turn, ú soon, þ **think**

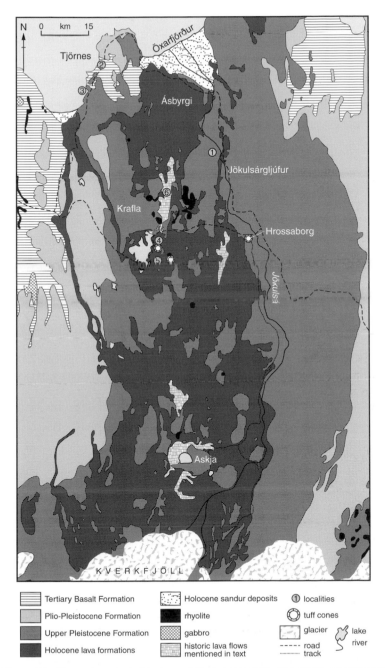

Figure 7.1 The main geological features of Northeast Iceland. Also shown are excursion routes and sites.

Figure 7.2 Geological cross section of Jökulsárgljúfur.

water that issues from cracks in the wall is in stark contrast to the brownish glacial water rushing forth at the bottom of the canyon.

Strictly speaking, the Jökulsárgljúfur is not entirely a canyon. It can be divided into three main parts. The southernmost part, 9 km long, is a clear canyon with near-vertical walls and carved out by fluvial erosion that follows fault lines trending north–south. Here the canyon walls comprise six lava flows and two sedimentary units (Fig. 7.2). The upper sedimentary unit occurs between lavas 3 and 4. Undercutting and collapse of lava 3 caused by erosion of this sedimentary unit resulted in the formation the Selfoss waterfall. The second sedimentary unit occurs at the foot in the deepest part of the canyon where two very thick lava flows (44 m in total) overlie it. Undercutting and collapse of these flows, induced by erosion of this lower unit, have formed Dettifoss, the greatest waterfall in Iceland in terms of volume discharge. The middle part of Jökulsárgljúfur is a 9 km-long U-shape valley with boulder and gravel terraces at different heights on both sides. The northernmost part of Jökulsárgljúfur is again a real canyon, 11.9 km long.

About 2 km to the northwest of the mouth of Jökulsárgljúfur is Ásbyrgi, a stunning impression into the land surface, which, according to the Viking mythology, was created when Sleipnir (Óðinn's horse) accidentally put one of his eight legs down on the ground. Ásbyrgi is an oval depression bounded by up to 100 m-high vertical walls on three sides, with an opening facing north. In its centre rises a cliff-bounded island.

The bulk of Jökulsárgljúfur, including Ásbyrgi, was formed in a jökulhlaup about 2500 years ago. The flood originated beneath the Vatnajökull icecap near the Bárðarbunga central volcano and was probably triggered by a subglacial eruption. It had a peak discharge of about 500 000 m³/s, and in total about 10 km³ of water was transported from the source to the sea at Öxarfjörður, a distance of 170 km.

The Sveinar graben system runs almost directly north–south for about 30 km and crosses the Jökulsá Canyon 4 km north of Dettifoss waterfall. West of the Jökulsá River, the cone row (volcanic fissure) Rauðuborgir runs along the centre of the graben system. On the east side of the canyon the graben system continues in the same direction, whereas the Randhólar cone row, which is a direct continuation of the volcanic fissure on the western side, sways to the northeast away from the graben. It is likely that the Sveinar graben was already in existence when the Rauðuborgir–Randhólar fissure eruption took place about 6000 years ago. It appears that the eruption started to the east of the river, and the fissure propagated to the south-southwest until it was captured by the graben and then continued within it towards the south. At Hljóðaklettar, just north of the Dettifoss waterfall, the canyon cuts through one of the cones of the Rauðuborgir–Randhólar fissures, exposing its interior and the top 100 m of the feeder dyke (Fig. 7.3). This outcrop provides a unique picture of a cone row in three dimensions.

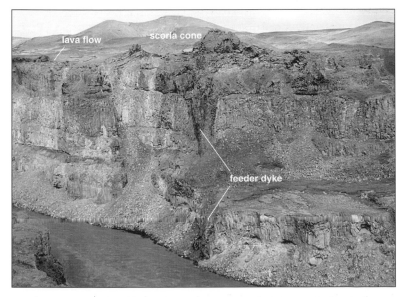

Figure 7.3 Section through the Rauðuborgir cone and its feeder dyke in the eastern wall of Jökulságljúfur.

Locality 7.2   Tjörnes sediment succession – transition from Pliocene to Pleistocene climate

At Tjörnes is a 1200 m-thick near-continuous stratigraphical section that records about 3 million years of Iceland's geological history, extending from the late Pliocene (at 4 million years) and into the Upper Pleistocene (0.7 million years). This remarkable succession provides an extensive record of how the fauna, flora and climate changed throughout Plio-Pleistocene times. The succession at Tjörnes features thick sequences of fluvial, estuarine and fossiliferous marine deposits, which are intercalated with several series of lava flows and some tillite beds. It is divided into three main sequences, which from oldest to youngest are the Tjörnes (proper), Furuvík and Breiðavík groups (Fig. 7.4).

The Tjörnes Group rests on an eroded lava pile of the Tertiary Basalt Formation, formed more than 8 million years ago, as can be observed in outcrop in the Kaldakvísl ravine on the western shores of the peninsula. The cliff sections along the coast offer excellent exposure through the group, where it can be followed in a continuous 6 km-long outcrop north to Höskuldsvík (Figs 7.4, 7.5a). The beds dip 5–10° to the northwest and are dislocated by faults in a few places. The cumulative stratigraphical thickness of the Tjörnes Group is about 500 m and it consists mainly of marine sediments rich in fossilized shells and snails (Fig. 7.5b). Intercalated with the marine beds are fluvial and lacustrine sediments, as well as occasional lignite seams.

The oldest sediments of the Tjörnes Group are about 4 million years old and the youngest about 2.5 million years old. It is divided into three sedimentary sequences on the basis of key reference fossils (i.e. mollusc species that distinguish each sequence). The oldest are the *Tapes* layers, then the *Mactra* layers, followed by the *Serripes* layers (Fig. 7.4).

Tapes   The *Tapes* layers, which outcrop between the Kaldakvísl and Reká streams, comprise a sediment sequence consisting of alternating shell-rich marine and lignite beds. Both are formed in a nearshore environment, the former in shallow coastal water and the latter above sea level in marshy areas adjacent to the shore. Today, the index animal, *Tapes*, lives no farther north than the North Sea, indicating much warmer seas around Iceland at the time of the sediments' formation (4–3.5 million years ago) than today.

Mactra   The *Mactra* layers that are found in the area between the streams Reká and Hallbjarnarstaðará take their name from an extinct mollusc species. This sequence is also characterized by alternations of shell-rich deposits and lignite seams, and the sediments were deposited in an environment similar to that of the *Tapes* layers. Some of the species found in the

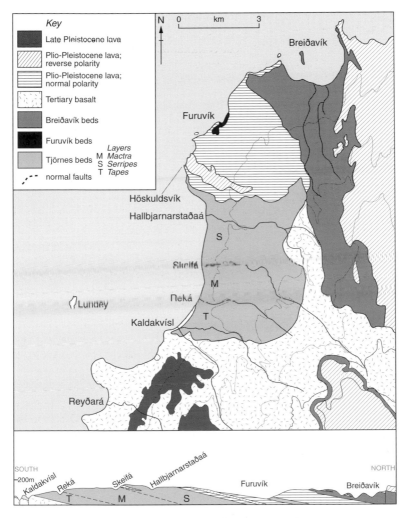

Figure 7.4 The Tjörnes sedimentary succession.

Mactra layers live today in much warmer seas, mainly in the region between the North Sea and the Canary Islands. Fossil pollen from the lignite seams shows that coniferous forest was growing on land surrounding the marshy areas. In addition to fir, spruce, pine and larch, broad-leaf trees such as oak and beech were growing in these forests, along with alder, birch and willow. Thus, both the fossil marine fauna and the terrestrial flora indicate a much warmer climate than at present. The winters were mild, with mean monthly temperatures above 0°C, and the waters around Iceland were at least 5°C warmer than today.

141

Figure 7.5 The Tjörnes sedimentary beds: **(a)** view of the *Tapes* layers in coastal cliffs at Tjörnes; **(b)** sedimentary bed rich in fossil shells.

Serripes The *Serripes* layers, which outcrop in the coastal banks between Hallbjarnarstaðará and Höskuldsvík, account for just over half of the thickness of the Tjörnes beds. The sequence consists almost entirely of marine sediments, with a few thin lignite seams towards the top. The lava that caps the *Serripes* sequence is 2.5 million years old. The fossil fauna at the base of the *Serripes* layers shows that most of the warm-water molluscs have disappeared from the waters around Iceland and are replaced by molluscs that prefer much colder seas. However, the sea temperature must have been somewhat higher than at present, because some warm-water species (e.g. *Solenensis*) were still living off the north coast of Iceland at that time. The most astonishing thing about this change in the fauna is that, out of 100 fossilized species of mollusc, about 25 per cent originated in the Pacific Ocean. These molluscs migrated from the Pacific across the Arctic into the Atlantic Ocean, when the sea flooded the Bering Strait for the first time. Most of the Pacific molluscs are cold-water species, including *Macoma* and *Serripes grönlandicus*, which is the index fossil for this sedimentary sequence. Therefore, the sudden cold-climate appearance of the fauna at the base of the *Serripes* layers may be a false alarm, because it probably does not indicate onset of sudden climatic cooling. It is more probably an artefact of the path the molluscs followed as they migrated from Pacific to Atlantic. The only molluscs that made it across were those that could survive in the much colder Arctic seas. On the other hand, the opening of the Bering Strait would have caused some changes in the ocean-current circulation and thus the climate. This change may have enhanced the cooling trend that was already in progress, because, shortly afterwards, the frigid winter of the Ice Age arrived in full force.

The Tjörnes and Breiðavík groups are separated by 250 m-thick series of basalt-lava flows that represents about half a million years of volcanic activity (2.5–2 million years ago). The Furuvík Group is in the middle of this lava series (Fig. 7.4) and is composed of two tillite beds with a cumulative thickness of about 40 m. These are the oldest glacial remains at Tjörnes and they show that the regional ice sheet extended to the northern shores of Iceland about 2.2 million years ago. This is taken to mark the onset of full glaciation in Iceland (see Fig. 1.19).

The Breiðavík Group rests on the lavas that cover the Furuvík deposits (Fig. 7.4) and comprises a 150 m-thick sequence of fossiliferous marine sediments alternating with tillite beds and occasional basalt lava. The Breiðavík Group was formed over 800 000 years ago (2.0–1.2 million years ago) and, in all, six tillite beds have been identified within the sequence, each representing a glacial stage. Most of the fossil fauna point towards ocean temperatures similar to that of the present, suggesting that the marine deposits were formed during interglacial stages. However, shells of *Portlandia arctica*, a mollusc that lived in then ice-cold waters of the Arctic, have been found on top of two tillite beds. They probably lived at the edge of the ice sheet at Breiðavík during the glacial stages. Six more tillite beds have been identified in the succession above the Breiðavík Group, bringing the total of glacial and interglacial stages recorded at Tjörnes to 14.

Locality 7.3   The Húsavík transform fault
The Húsavík fault is the principal fault of the Tjörnes Fracture Zone and, as the name implies, it cuts through the town of Húsavík (see Fig. 7.1). Off shore, this fault forms a valley 5–10 km wide and 3–4 km deep that trends towards the sea stack, Kolbeinsey. The Húsavík fault is characterized by right-lateral fault movement (i.e. standing on one side of the fault, the block on the other side has been displaced to the right). It has been active for about 9 million years, during which time the cumulative displacement along the fault has been 60 km.

The fault planes of these fractures are seen on land at the Tjörnes Peninsula as linear northwest-trending structures dissecting the bedrock. One of these fractures passes right through the town of Húsavík and is the very source of its existence, because here the displacement of rock masses along the fault has formed a perfect natural harbour. The total displacement on the fault over the past million years is in the order of several kilometres. This fault is best viewed from the top of Mt Húsavíkurfjall, which rises above the town. The fault is easily seen from there as it comes ashore at the harbour and passes through the centre of Húsavík and onwards towards the Þeystareykir rift segment southeast of Húsavík.

Krafla–Mývatn
A paradise of volcanology

The Mývatn area is not only an Eldorado for volcanology enthusiasts and bird watchers, but also has much appeal for those interested in landscape evolution, glacial geology and settlement geography. It is truly an extraordinary natural laboratory, because it features almost all aspects of the natural sciences within an area of no more than 15 km^2.

The area takes its name from Lake Mývatn, the third largest lake in Iceland (38 km^2), and is situated on the edge of the Krafla volcanic system. The lake is 227 m above sea level and its maximum depth is only 4 m. It sits in a shallow depression, which to the west is bordered by smooth ridges composed of Upper Pleistocene basalts and to the east and southeast by móberg ridges and three large table mountains, Sellandafjall, Bláfjall and Burfell. To the north is the basalt caldera volcano, Krafla, which has produced many eruptions throughout the Holocene. The most recent are the so-called Mývatn Fires (1724–9) and Krafla Fires (1975–84).

The Krafla volcanic system, its fissures and volcanic architecture
The Krafla volcanic system consists of a basalt caldera volcano and a elongate fissure swarm (see Fig. 1.5 and Table 1.4). The Krafla fissure swarm

is 10 km wide and 100 km long; it extends from the table mountain, Sellandafjall, in the south and into the sea at Öxarfjörður in the north. The Krafla central volcano is more than 200 000 years old and consists of a 10 km-wide caldera flanked by a broad and low-lying shield, chiefly composed of basalt-lava flows (see Fig. 7.1). Thus, the volcano is built up from both subglacial and subaerial eruptions. The caldera itself was formed in a large explosive eruption that produced the mixed basalt–rhyolite volcanics of Mt Hágöng during the most recent interglacial (> 100 000 years ago). Along outer margins of the caldera are two rhyolite lava domes, Hlíðarfjall and Jörundur, which were formed in subglacial eruptions some 20 000–30 000 years ago; they are good representatives of rhyolite table mountains. Within the caldera is a major high-temperature geothermal area, which today is harvested for electricity. Another high-temperature geothermal field, utilized by the diatom-mud processing plant, is situated out on the fissure swarm at the Námafjall Ridge.

The postglacial history of the Mývatn area
During the Younger Dryas stage, the northern margins of the ice sheet were situated along the east–west trending end moraines just north of the Reykjahlíð farm (Fig. 7.6). In front (i.e. north) of the Reykjahlíð moraines, glacial rivers formed a small sandur plain that extends north to Mt Hlíðarfjall. Later, this sandur plain has been split up by dislocations (faulting), forming clear exposure through the gravelly glaciofluvial deposits.

After the Younger Dryas glacier receded from the area, subaerial volcanism set in on a larger scale. The Holocene volcanism began with the hydromagmatic eruption that formed the tuff cone Lúdent (Fig. 7.6). This eruption episode continued with lava outpourings from volcanic fissures in the Lúdent area and also from short fissures west of the Námafjall Ridge. Lava flows from these fissures are visible just east of the Reykjahlíð farm and on the islet of Slúttnes. All of the lava flows from this period predate the rhyolite fall deposit (H5) produced Hekla about 6600 years ago. Near the end of this episode the crater Hraunbunga produced the dacite lava that flanks the Lúdent tuff cone, after which volcanism in the area went quiet for more than 3000 years.

About 3800 years ago, the volcanoes began to rumble again. The first to make its mark was the lava shield Ketildyngja, about 25 km southeast of Mývatn, which was the first effusive eruption to have a major impact on the landscape evolution in the Mývatn area. This eruption is the source of the Older Laxárdalur lava flow, which entered the Mývatn area through the narrow Selhjallagil gorge (Fig. 7.6). As the lava spread across the basin, it dammed up the outflow from the Mývatn depression and formed the first Lake Mývatn. The lava field known as Grænavatnsbruni is part of this

145

Figure 7.6 The Mývatn area.

enormous flow, which advanced down the Laxárdalur and Aðaldalur valleys, reaching Skjálfandi Bay, 50 km north of Mývatn (see Fig. 7.1).

About 3000 years ago the Mývatn area was covered by rhyolitic tephra fall from the Hekla volcano in South Iceland. This is the so-called H3 tephra layer, which appears as a 5–6 cm-thick white band in the soil profiles around Mývatn. A little later (2800 years ago) a new eruption episode began in the Mývatn area, with the birth of the near-perfectly shaped tuff cone, Hverfjall. This eruption was followed by a series of fissure eruptions,

Figure 7.7 A rootless cone on the shore of Lake Mývatn.

including the spectacular Þrengslaborgir–Lúdentsborgir cone row, which formed the Younger Laxárdalur lava flow about 2000 years ago (Fig. 7.6). Like its predecessor, this lava was a major contributor to sculpturing the landscape in the Mývatn area and is mainly responsible for the scenery as we see it today. It shaped the lake to its present size and formed the maze-like landscape of Dimmuborgir. The rootless cones, which border the lake shores and form its many islets, were also formed in this eruption when the lava circumvented the ancestral Lake Mývatn. They are truly the trademarks of the scenery at Mývatn and their often symmetrical outlines make them real miracles of nature (Fig. 7.7). Several small fissure eruptions followed the Þrengslaborgir–Lúdentsborgir eruption in the area north of Hverfjall and of those the most conspicuous one is the Svörtuborgir ("black fortress") cone row.

Yet again, the volcanoes in the Mývatn area fell silent, this time for just over a thousand years. On 17 May 1724 the silence was broken and the volcanoes began to rumble again. A new period of volcanic activity had begun within the Krafla volcanic system, first with the volcanotectonic episode that became known as the Mývatn Fires and another one about 250 years later, namely the infamous 1975–84 Krafla Fires.

Locality 39   Hverfjall tuff ring

Hverfjall ("crater mountain") is an almost perfectly circular tuff cone formed in a hydromagmatic eruption through the ancestral Lake Mývatn about 2700 years ago. The cone was produced by the main phase of the so-called Hverfjall Fires and is the largest structure on a 2 km-long volcanic fissure, 90–150 m high and about 1000 m in diameter (Fig. 7.6). It consists of many tephra beds dipping 15–35° in all directions from the crater. These beds are formed in two ways: as fallout from the eruption column and as pyroclastic surges transported by laterally moving collar-like clouds.

Away from the cone, the tephra fall was heaviest to the south and northeast of the cone (Fig. 7.8a), but the pyroclastic surges were probably dispersed fairly evenly around the cone. North of Hverfjall are excellent outcrops into the tephra deposit and here the sequence consists of alternating pyroclastic-surge and tephra-fall deposits. Some pyroclastic surges were very wet, because their deposits are plastered, layer upon layer, onto tree trunks. Some trunks were bent to the ground by the extra weight of the tephra (Fig. 7.8b). Other surges were hotter and drier, and flowed faster, forming dunes and beautiful cross-bedded deposits in the process (Fig. 7.8c). These surges travelled more than 3 km away from the vent.

Jarðbaðshólar is the group of the northernmost and largest (40 m-high) cones on a 2 km-long volcanic fissure situated immediately to the north of

Figure 7.8 **(a)** The thickness and dispersal of the Hverfjall tephra fall deposit in the Mývatn area. **(b and c, opposite)**

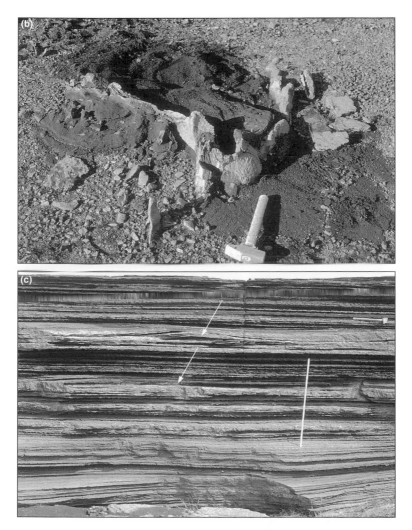

Figure 7.8 **(b)** A tree mould in the Hverfjall tephra layer, showing how the tree trunk collapsed from the weight of the tephra plastered onto it by wet pyroclastic surges. The handle of the hammer points in the direction of flow. **(c)** Cross-bedded deposits (arrows) within the Hverfjall tephra layer formed by dry pyroclastic surges (the ruler is 30 cm long).

Hverfjall (Fig. 7.6). The Jarðbaðshólar fissure produced the pahoehoe and rubbly pahoehoe lava that covers the ground from the fissures to the eastern shores of Lake Mývatn. In the later stages of the eruption, the lava flowed away from the northern cone in a channel that extends almost half way to Lake Mývatn. The lava that emerged from these channels formed the rubbly lava that covers the area east of Vogar. The Jarðbaðshólar lava covers about 8 km^2 and its volume is 0.05 km^3. The Jarðbaðshólar eruption occurred shortly after activity ceased at the Hverfjall vents and represents a separate episode in the Hverfjall Fires.

Locality 7.5 The Þrengslaborgir eruption and origin of Lake Mývatn
The Younger Laxárdalur lava flow was formed by an effusive eruption on the Lúdentsborgir–Þrengslaborgir cone row about 2300 years ago. The cone row runs from the Lúdent tuff cone some 12 km to the south and then shifts to the west onto the Borgir cone row (Fig. 7.6). The fissure's total length is about 16.5 km. Most of the erupted material came up on the central part of the Þrengslaborgir fissure and fed the lava that formed the Dimmuborgir complex and the rootless cones around Lake Mývatn. The lava continued its advance down the Laxárdalur Valley and all the way to the north coast. The total volume of the Younger Laxárdalur is 2–3 km^3.

Dimmuborgir is a roughly circular shield-like structure about 2 km in diameter. Forming a natural labyrinth of drained lava tubes, it originated as a secondary lava pond within the Þrengslaborgir lava field, formed when the advance of the flow was hampered by the many rootless eruptions that occurred when the lava entered the ancestral Lake Mývatn. This is evident in the walls inside the labyrinth, which were constructed by many thin overbank flows spilling out of the pond. When the lava had established dry pathways across the lake, the pond was drained rapidly, leaving behind caved-in structures and ramparts with relict lava tubes and the filled pathways of ancestral lava rivers.

Locality 7.6 The 1724–9 Mývatn Fires and 1975–84 Krafla Fires
Two major volcanotectonic episodes have occurred in the Krafla volcanic system in the past 250 years: the 1724–9 Mývatn Fires and the 1975–84 Krafla Fires. This activity, especially that of the Krafla Fires, has provided geologists with valuable insights into the mechanics of volcanic systems in Iceland and is therefore summarized here in some detail.

The Mývatn Fires The volcanotectonic episode referred to as the Mývatn Fires lasted for about five years, from 1724 to 1729. It was characterized by major rifting and faulting on the Krafla volcanic system and was accompanied by a series of explosive and effusive eruptions (Fig. 7.9). The episode began on 17 May 1724 with earthquakes and a hydromagmatic eruption, which formed the Víti crater on the southwest slopes of Mt Krafla. Intense earthquakes and large-scale faulting in the area between the Krafla volcano and Lake Mývatn followed this eruption. In January 1726, strong earthquakes were felt again and small explosive eruptions occurred at the Leirhnjúkur hill west of Mt Krafla. Some activity was noted

Figure 7.9 The vents and lavas from the 1724–29 Mývatn Fires.

at Bjarnarflag in April that year. On 21 August 1727 a fissure (Skeifa) opened at the northern end of the Krafla caldera, sending lava northeast towards Gæsafjöll. Eight months later, on 28 April 1728, rifting resumed, with lava issued from the Rauðkollur fissure segment in the Krafla caldera and from short fissures in Hrossadalur, a short distance east of the Reykjahlíð parsonage (Fig. 7.9).

Disturbance resumed in December. On 18 December a large amount of lava belched from the Hófur fissure and from a short fissure at Bjarnarflag, just east of the Reykjahlíð parsonage. Jón Sæmundsson, the vicar at Reykjahlíð described this volcanic outburst in Bjarnarflag as follows:

> In some places the fire had greater action with huge and great flames, and continuously threw up all around itself glowing rocks and stones, which are so wet [fluid] that they, still glowing, flowed like melted copper. This ejection of glowing rocks forms large crater-like hills or rocky hills [where the fire emerges], which are hollow inside with circular rims. Inside these source pits of the fires were an intensive and strong *inflammatio ignis* with such a strong storm that the flames were thrown high up into the sky. The same fire threw up continuously glowing and wet rocks that when they hit the ground, they flowed like running water and burned up everything in their way.

It is obvious that Jón Sæmundsson is describing fire fountains, and his depiction of glowing and "wet" rocks thrown up into the air and flowing like water after hitting the ground is the oldest known description of a fountain-fed lava flow. However, Krafla was not done and another eruption started on 30 June 1729. This time a fissure opened at Hítarhóll just south of Leirhnjúkur, sending a large volume of lava down towards Reykjahlíð and all the way into Lake Mývatn (Fig. 7.9). This flow ruined four farms, including the Reykjahlíð farm, and forced the vicar and his family to seek shelter at Skútustaðir. However, the lava did not manage to destroy the Reykjahlíð church, which stood on a small hillock and still does, and now is completely surrounded by lava. Seventeen years later, in July 1746, a small explosive eruption occurred at Leirhnjúkur, suggesting that this volcanotectonic episode at Krafla may have lasted for up to 22 years. The cumulative length of the volcanic fissure formed in the Mývatn Fires is about 13 km

The Krafla Fires The volcanotectonic episode of the Krafla Fires began in December 1975 and lasted until September 1984. The first sign of renewed activity within the Krafla volcanic system was an increase in earthquake activity in June 1975, gradually building up to earthquakes of magnitude 4 on the Richter scale. The first eruption broke out from a short fissure at Leirhnjúkur, producing a small lava flow and a few explosions (Fig. 7.10). Initially, the earthquake activity was confined to the region below the caldera, but it spread out quickly along the fissure swarm to the north coast at Öxarfjörður and south towards Lake Mývatn. This rifting was associated with a 1.5 m widening of the fissure swarm, and a 2 m-deep graben was formed along the central part of the fissure system. The eruption was followed by an immediate subsidence within the caldera floor, forming a bowl-shape depression. This was followed by a gradual inflation of the caldera floor over the next few months. This pattern of gradual inflation and sudden deflation recurred several times in the following years. Each cycle began with inflation of the caldera floor at a rate of 7–10 mm/day. When the inflation reached a critical level, the earthquake activity increased steadily until another sudden deflation event occurred. This is known as a rifting event. During the Krafla Fires, the rifting began by failure of the crust within the caldera and migrated from there along the fissure swarm, as indicated by the outward-propagating earthquake swarms. Magma flowed into and filled the open cracks (see Fig. 1.6b) and in some cases found a pathway to the surface to produce an eruption.

In total, 21 rifting events occurred during the 1975–84 Krafla volcanotectonic episode. The first four rifting events occurred in October 1976, January, April and September 1977, but only the last two deflation events resulted in an eruption at the surface. Other notable rifting events occurred, for example, in March, July and October 1980, November 1981 and September 1984. Many of the later deflation events resulted in fairly large lava eruptions (Fig. 7.10). After the 1984 event, inflation resumed within the caldera until it ceased in early 1985. Minor, intermittent inflation/deflation cycles were recorded between 1986 and 1989; since then the caldera floor has been subsiding at a slow but steady rate.

The total widening of the Krafla fissure swarm during the nine-year volcanotectonic episode was in the order of 900 cm and the widening during the 1724–29 Mývatn Fires was of the same magnitude. As you might recall, the long-term average spreading rate of the plate boundary in Iceland is 1.8 cm/year. Thus, through rifting in two relatively short volcanotectonic episodes spanning 15 years in total, the volcanic system has made up for the previous 1000 years, which were marked by little or no spreading on this particular segment of the plate boundary. This evidence shows that the rifting and spreading along the plate boundary are not a continuous process, but are periodic and at any time confined to episodes, such as the Mývatn and Krafla Fires, on individual volcanic systems.

153

Figure 7.10 The developments of vents and lavas during the 1975–84 Krafla Fires.

Chapter 8

The north and northwest

General overview

North and Northwest Iceland are composed entirely of rocks belonging to the Tertiary Basalt Formation, except for the outlier and extinct Skagi volcanic belt which was formed during the Plio-Pleistocene time. Similar to East Iceland, these regions are characterized by rugged mountains and deep fjords, a landscape of extensive glacial erosion.

The age of the Tertiary Basalt Formation in North Iceland ranges from about 12 million years in the deepest part of the stratigraphical column at the northern tip of Tröllaskagi to about 7 million years in the region from Blönduós across the Central Highlands to the bottom of Eyjafjarðardalur. At the Skagi Peninsula the much younger (<3 million years) Skagi volcanic belt is superimposed on the eroded Tertiary Basalt Formation (Fig. 8.1). Even older rocks are exposed at the Vestfirðir Peninsula, where some, 16 million years old, occur deep in the stratigraphical column in the far northwest. The succession becomes progressively younger towards the southeast and at the neck of the peninsula; the age is about 8 million years

The construction of the Tertiary Basalt Formation in North and Northwest Iceland was identical to that of East Iceland, and altogether 23 extinct volcanic systems have been identified in the succession. It also contains six major sedimentary horizons (Fig. 8.1), which are, from west to east, the Selárdalur–Botn (15 million years old), Saurbær (13 million), Brjánslækur (12–11 million), Bjarkarlundur (11–10 million), Húsavíkurkleif–Tröllatunga (10–9 million), and Tindar–Mókollsdalur (8 million). The sediments represent fluvial, lake or peat-bog (lignite) deposits and are all formed on land.

All of these sedimentary horizons contain fossils or fossil imprints of the plants that lived at the time. For example, the 15 million-year-old Selárdalur–Botn horizon, which extends from Selárdalur to Kögur (Fig. 8.1), contains 20 m-thick lake sediments with leaf imprints and pollen (seeds) of trees such as vine, walnut, elm, beech, dawn redwood and pine. This fossil fauna most resembles the eastern deciduous forest of North

á house, æ **eye,** au feuille, ð **that,** ei day, é yet, hv **quick,** í seen, ö turn, ú soon, þ **think**

▤ Tertiary Basalt Formation	⬚ Holocene sandur deposits
▦ Plio-Pleistocene Formation	⬛ rhyolite
▨ Upper Pleistocene Formation	▩ gabbro
⬛ Holocene lava formations	🗺 lake

0   km   50   N

Figure 8.1 The main geological features of North and Northwest Iceland. Also shown are excursion routes and sites.

America, showing that at this time the climate of Iceland was similar to that of the southeastern USA today. In general, fossil fauna indicate a gradual and steady cooling of the climate from 16 to 7 million years ago, which is consistent with similar evidence elsewhere. Each of these sedimentary horizons can be followed for tens of kilometres and they appear to mark extended periods of subdued volcanic activity. A break in lava production allowed for thick sedimentary sequences to accumulate without interruption. The apparent 1 million-year cyclicity in their occurrence is interesting and may be recording the life cycle of the volcanic systems that constructed the bulk of the succession. There are also some significant structural changes to the Tertiary Basalt Formation across the Selardalur– Botn sedimentary horizon. Above the horizon the formation is characterized by a shallow regional dip (5-10°) to the southeast, whereas below it the dip is about 5° to the west (see Fig. 1.2). This reversal in the regional dip implies different source regions for the successions above and below the Selardalur horizon. The southeast dip of the overlying succession is consistent

156

with an origin within the now-extinct Snafellsnes–Vatnsdalur Volcanic Zone, whereas the westward dip of the succession below the horizon implies origin within a volcanic zone that is now buried in the sea somewhere off the northwest coast of Iceland.

There is one excursion in this region, which follows Highway 1 across North Iceland from Fnjóskadalur to Hrútarfjörður (Fig. 8.1).

Fnjóskadalur to Skagi

Locality 8.1 Fnjóskadalur valley and the ice-dammed lake sediments
The Bleiksmýrardalur Valley and its northward extension, the Fnjóskadalur Valley, form the longest valley system in Iceland, 150 km in total. Situated within the Tertiary Basalt Formation, it was originally formed by fluvial erosion during the late Tertiary period and was later carved out by the Ice Age glaciers. Scattered outcrops into young Upper Pleistocene lavas show that lava flows came down this valley system during one of the interglacial stages and at least one advanced all the way into Eyjafjörður (Fig. 8.1). During the waning stages of the Weichselian glaciation (i.e. Allerød stage), the Fnjóskadalur valley was occupied by an ice-dammed lake, which was kept up by the valley glacier that filled Eyjafjörður (Fig. 8.2). The gravel terraces that circumscribe the lower valley slopes were formed in this lake. The so-called Skógar tephra, a pale pumice-fall deposit of rhyolite composition, is found in these terraces. This tephra layer was formed by a large explosive Plinian eruption in the Mýrdalsjökull volcano some 10 600 years ago, which formed the Sólheimar ignimbrite. The eruption plumes also dispersed ash over much of the North Atlantic and Scandinavia, to a distance of more than 1500 km from its source.

On the way. The extinct Öxnadalur central volcano, which was active 8–9 million years ago, is located southwest of the town of Akureyri, where Highway 1 climbs out of the Öxnadalur Valley. In its time the volcano rose to about 1000 m above the surroundings and was capped by a 6 km-wide and 1 km-deep caldera. An exceptional exposure of the eastern caldera wall is clearly visible in the top cliff face of Mt Landafjall, right above the Þverá farm. Following the distinctively layered basalt-lava series that forms the outer slope of the volcano from the eastern end of Mt Landafjall, it is suddenly terminated by a near-vertical caldera wall near the western end of the mountain. Here, irregularly bedded sediments and volcanics fill in the caldera. The volcanics consist of cube-jointed basalt flows, pillow lavas, and móberg tuffs and breccias, indicating that at some stage the caldera contained a lake.

The pile of debris that forms the hills at the foot of the mountains on the

Figure 8.2 The condition during Allerød Stage when the ice-dammed lake filled the Fnjóska-dalur valley. Also shown are the positions of the main valley glaciers during the Allerød stage.

opposite side of the valley is called Hraun and was formed as a rock glacier during early postglacial times. The rock glacier blocked the small valley above and a small lake formed in the depression behind its terminus. A little farther west, near the deserted Bakkasel farm at the bottom of Öxna-dalur, a large rock called Lurkasteinn stands proud to the south of the road. It is mentioned in one of the Icelandic sagas (i.e. Þórðar sögu hreðu) as the scene of a duel between Þórður the squabbler and Sörli the sturdy one, which resulted in Sörli's death.

Locality 8.2 Kotagil

The road across Öxnadalsheiði pass descends into Norðurárdalur, a narrow glacially carved valley with an almost perfect U-shape cross section. On the north side of Norðurárdalur, the small tributary stream Kotá flows through an impressive gorge, which has been cut into the Tertiary Basalt Formation by recurrent but infrequent flash floods.

In the west wall of the gorge near its base, about 20–30 m from its mouth, there are moulds of ancient tree trunks at the contact between two basalt-lava flows (Fig. 8.3). The presence of such tree moulds has long been noted in Tertiary lava flows in Iceland and can be distinguished from other voids by a boxwork pattern on their inner surface, or by the branching form of the mould itself. In pahoehoe lavas, tree moulds are typically found near the base or top of a flow lobe and, as in Kotagil, there is never any soil or sediment between the two lobes. Therefore, it is unlikely that the trees were standing on the lower lava when they were alive and growing. So, how did the tree trunks assume a position where they were buried and burned by the lava to form the moulds? We know from observations in Hawaii that coconut palms are burned and toppled onto the surface of inflating pahoehoe flows. The trees are preserved where subsequent lava lobes cover the log before it decays. The Kotagil and tree moulds in Icelandic lavas were most probably formed in a similar way.

On the way. As you follow Highway 1 along the southern slopes of Bólstaðarhlíðarfjall, just before entering the Langidalur valley en route to Blönduós, a keen observer might notice the many dislocations visible within the westward-dipping strata. These are classic examples of normal faults, where one side has been displaced downwards relative to the other.

Locality 8.3 Vatnsdalshólar landslide formation

Vatnsdalshólar is a chaotic cluster of hummocks and hills that stretch across the mouth of the Vatnsdalur Valley. These mounds and hummocks are made entirely of assorted rock debris and are one of the best examples in Iceland of deposits produced by a rock avalanche. Here, the flow-like shape and the hummocky surface morphology that characterize deposits of this type of are perfectly preserved. The avalanche that formed Vatnsdalshólar occurred more than 7000 years ago. The scarp left behind by the collapsing rock mass can still be seen in the western slope of Mt Vatnsdalsfjall at about 900 m above sea level. As the avalanche hurtled down the slope, it picked up momentum and ran up the opposing slope to a height of about 70 m. It also spread debris up to 4 km away from its source and covered an area of about 15 km².

Two smaller rock avalanches have descended from Mt Vatnsdalsfjall in historical times. The first one cascaded down the mountain slopes in 1545

Figure 8.3 Tree moulds at the point of contact between two pahoehoe lobes at Kotagil in North Iceland

and ran over the Skíðastaðir farm, killing 14 people. The small lake, Hnausatjörn, just south of the highway, is formed in the debris from this avalanche. The second one occurred in 1720 and surged down the slopes just south of Hnausatjörn. The debris dammed the river and formed the larger lake, Flóðið. Six people were killed in this event.

160

Locality 8.4 Vatnsnes syncline and the Skagi volcanic succession
In approaching the Vatnsnes Peninsula, note that the regional dip of the strata has reversed. At Mt Vatnsdalsfjall and other mountains to the east, the strata dip at 5–10° west or west-northwest, whereas at Vatnsnes and the mountains farther to the west the strata generally dip to the east. The strata with these opposing dips converge on the east side of Vatnsnes, defining a north-trending axis of a syncline (see Fig. 1.2). This is the Vatnsnes syncline, which represents an old rift zone that became inactive about 5 million years ago and is the precursor to the currently active North Volcanic Zone. Prior to 5 million years ago, the northern limb of the actively spreading plate boundary across Iceland was situated at the Vatnsnes rift zone. The three extinct central volcanoes that stretch from Mt Vatnsdalsfjall across to Vatnsnes were formed by volcanism on this rift zone. The much younger (less than 3 million years old) Skagi volcanic succession (Fig. 8.1) was formed in a rejuvenation stage on this old rift zone – similar to recent volcanics on the Snæfellsnes Volcanic Belt.

When crossing the Holtavörðuheiði mountain road on a bright day, a large steep-sided mountain capped by a cupola-shape glacier appears on the horizon to the southwest. This is Eiríksjökull (1675 m), the largest table mountain in Iceland. The basal diameter of the mountain is about 10 km and it rises more than 1000 m above its surroundings. The lowest 350 m consists of móberg tuffs and breccias, which are capped by a lava shield 750 m thick. It was presumably formed in a single subglacial eruption some time during the most recent glacial stage.

Chapter 9

The west

The Snæfellsnes Peninsula is about 80 km long but only 10–30 km wide (Fig. 9.1). A 700–1000 m-high mountain chain forms the backbone of the peninsula, which at its tip is crowned by the Snæfellsjökull stratovolcano.

Snæfellsnes–Borgarfjörður
Volcanic zones and rift jumps

Snæfellsnes Volcanic Belt – the mantle-plume trail
The mountain chain that forms the crest of the Snæfellsnes Volcanic Belt is the product of outlier volcanism that began about a million years ago along the trail of the mantle plume. These young volcanics rest unconformably on the 1.2 million-year-old Búlandshöfði sedimentary horizon, which in turn rests on the much older and eroded bedrock of the Tertiary Basalt Formation. The Snæfellsnes Volcanic Belt comprises three volcanic systems trending west-northwest, which exhibit a subparallel (en echelon) arrangement along the peninsula. These are, from east to west, the Ljósufjöll, Lýsuskarð and Snæfellsjökull volcanic systems (see Fig. 1.5). The small cluster of Upper Pleistocene volcanics at Snjófjöll may represent the fourth system of this belt.

The 90 km-long and 20 km-wide Ljósufjöll system is the largest of the three systems and has been active for at least 0.7 million years. Its centre is at the multi-coloured Ljósufjöll mountain complex, which consists of andesite to rhyolite volcanics, mostly formed by subglacial eruption during the Upper Pleistocene. During the Holocene, the activity has been characterized by the basaltic volcanism that often extends across the system, from the fissures of Berserkjahraun in the west to the Grábrók scoria cone in the east. In all, 23 basalt eruptions have occurred in the system in the past 10 000 years, including one historical eruption and the well known spatter-cone volcano, Eldborg á Mýrum (see locality 9.3). The latter is the source of the term "eldborg", commonly used in Iceland for spatter cones.

The Lýsuskarð system is located in the Central Highlands of the

á house, æ eye, au feuille, ð that, ei day, é yet, hv quick, í seen, ö turn, ú soon, þ think

Figure 9.1   The main geological features of West Iceland, centered on the Snæfellsnes Peninsula. Also shown are excursion routes and sites.

peninsula and is the smallest one, 30 km long and 5 km wide. The core of this system is a central volcano that began to erupt about 2 million years ago, but was most productive after 1 million years. Only two eruptions are known within the system in the past 10 000 years, both basalt-lava eruptions, one forming magnificent lava falls that cover the southern slopes of the highlands near Lýsuhóll.

The third volcanic system, Snæfellsjökull, which is named after the infamous stratovolcano that sits on the western tip of the peninsula, is about 30 km long and has been active for just over 700 000 years. Although the Snæfellsjökull system has been very active during the Holocene, with a tally of more than 25 effusive basaltic and 3 explosive Plinian eruptions, none have occurred in the past 1100 years, the time of human settlement in Iceland (see also locality 9.2).

The Tertiary geology: extinct rift zone and rift jump
The Tertiary geology of western Iceland tells an intriguing story of a rift jump, because it dictates how plate boundaries were shifted from one volcanic zone to another. Unfortunately, the geology is complex and thus

cannot be unravelled without involving demanding concepts, so please bear with us.

The youngest rocks of the Tertiary Basalt Formation at Snæfellsnes are about 6 million years old. They occur along the axis of the Snæfellsnes syncline, which stretches obliquely across the peninsula from Garðar to Álftafjörður, where it turns eastwards across Hvamsfjörður (see Fig. 9.1). On either side the strata dip towards the syncline axis, and the rocks become progressively older away from it. Northwards, the Tertiary succession can be followed without a major break all the way across the Vestfirðir Peninsula or back to rocks that were formed about 16 million years ago. To the south they can be followed across the Mýrar region before the Tertiary rocks disappear beneath the much younger products of the West Volcanic Zone on the south side of Borgarfjörður (see Fig. 9.1).

The Snæfellsnes syncline is an ancient rift zone and was characterized by volcanism identical to that found in the currently active volcanic zones; it hosted at least four active volcanic systems in its prime. This rift zone was an active spreading centre until 6 million years ago, when the plate boundary shifted to its current position within the West Volcanic Zone. It is a continuation of the extinct Vatnsnes rift zone and it represents the southern limb of the plate boundary across Iceland as it was prior to 6 million years ago.

Along a zone extending northwest from Borgarnes to Lake Langavatn, the dip of the Tertiary strata reverses again (see Fig. 9.1). North of this zone the strata dip to the northwest, whereas on the south side they dip to the southeast. This is the Borgarfjörður anticline (see locality 9.4). The oldest rocks in Borgarfjörður are exposed in the axis of the anticline and are 13 million years old. Away from the anticline axis the rocks become younger: 6–8 million years old at Snæfellsnes and Hítardalur, and about 12 million years old where the Tertiary basalts disappear under the younger products of the West Volcanic Zone to the south.

In the Tertiary strata on either side of the Snæfellsnes syncline are thick conspicuous sedimentary units, the Tindar and Hreðavatn horizons (see Fig. 9.1), which were formed about 8 million years ago during a period when there was a lull in volcanic activity. The Tindar horizon occurs within a succession without major time breaks. The story is quite different for the Hreðavatn horizon.

The Hreðavatn sedimentary horizon can be followed from Hítarvatn east across to Hreðavatn and then southwest to Mt Hafnarfjall. At Hítarvatn, the locality closest to the Snæfellsnes syncline, the age difference between the lavas on either side of the Hreðavatn horizon is very small (less than 0.3 million years). Following the Hreðavatn horizon to the south and southwest, the age difference between the lavas below and above increases steadily. At Hreðavatn the difference is about 5 million years, the underlying lavas being 12 million years old, whereas the ones on top are 6.5 million years old. At Mt Hafnarfjall the difference is even greater at 7 million years. This age sequence tells us two things.

First, the age progression from Hítarvatn to Hreðavatn and Mt Hafnarfjall shows that the Tertiary volcanics become progressively older across the Mýrar region. This is to be expected if they were formed by volcanism in the now extinct Snæfellsnes rift zone and subsequently moved from it by spreading to the southeast. By the same token, the equal age of the lavas at Hreðavatn and Mt Hafnarfjall is expected because they are at the same distance from the spreading axis. Thus, moving southwards, the Hreðavatn sediments were blanketing progressively older and more eroded lava sequences. Consequently, away from the Hítardalur locality, the contact between the Hreðavatn sediments and the underlying Tertiary lavas is an unconformity, because part of the Tertiary succession is missing.

Secondly, the volcanic formations immediately above the sedimentary horizon become gradually younger from Hítarvatn across to Hreðavatn and Mt Hafnarfjall, indicating that 7 million years ago volcanism had begun in the area to the east of Hreðavatn and gradually spread to the southwest over the next 2 million years. At the same time the activity was gradually declining in the Snæfellsnes rift zone and the last volcano died out about 6 million years ago. This sequence clearly indicates that the centre of volcanism and spreading was shifting from the Snæfellsnes rift zone to the current West Volcanic Zone (see Fig. 1.3), a rift-zone jump that took at least 2 million years to complete. The Borgarfjörður anticline was formed as a result of this shift. First, the strata were tilted to the northwest because of loading of the crust by the volcanism at the Snæfellsnes rift zone. Later, the part south of the anticline axis was tilted to the southwest by loading within the West Volcanic Zone.

At present, this same process was in progress between the active West and East Volcanic Zones. The northwesterly drift of the North America/ Eurasia plate boundary relative to the Iceland mantle plume is the driving force behind these rift jumps. The history of these plate-boundary shifts and evolution of the rift zones in Iceland is illustrated in Figure 9.2.

Locality 9.1 Búlandshöfði interglacial sediment sequence
Towards the base of the mountains on the north side of the Snæfellsnes Peninsula is a 20–50 m-thick sedimentary sequence that can be followed for 30 km from Grundarfjörður to Ólafsvík. Good cross-sectional views through the sediments are found in outcrops at Mt Stöð and Mt Búlands-höfði, from which the sedimentary sequence takes its name. The Búlands-höfði sediments and overlying volcanic formations record two glacial

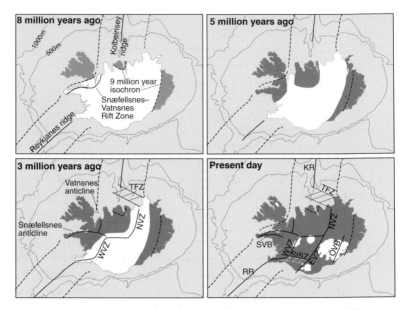

Figure 9.2 Evolution of rifts and volcanic zones in Iceland during the past 16 million years.

stages and one interglacial stage that occurred 0.9–1.2 million years ago.

At Stöð and Búlandshöfði the lavas of the Tertiary Basalt Formation rise to 130 m above sea level. The upper surface of the lava pile is striated, showing that the Tertiary Formation has been scraped and eroded by an advancing glacier. The tillite bed mixed with marine sediments rests unconformably on this surface. Mollusc fossils (e.g. *Protlandia arctica*) have been found in these marine sediments, which indicate an ice-cold sea. This sequence represents the first glacial stage recorded by the Búlandshöfði sequence. On top of these cold-sea sediments are clay-rich marine deposits containing fossils of molluscs that live in the ocean around Iceland today (i.e. *Arctica islandica*, *Mytilus edulis* and *Nucella lapillus*). These fossils show that the sea temperature and the climate had become much warmer when these sediments were laid down. The ice sheet of the previous glacial stage had receded and the region now enjoyed the warmer climate of an interglacial stage. These warm marine sediments are missing in the Mt Stöð section. In their place is a unit of fine-grain sandstone forming a delta built into a lake. This sandstone contains many leaf imprints and pollen grains of alder, willow and birch: the land was fully vegetated. Shortly afterwards the Búlandshöfði sediments were covered by a series of lava flows formed by fissure eruption within the rejuvenated Snæfellsnes volcanic zone about 1.1 million years ago. The upper surface of this lava sequence is polished and striated, showing that the ice sheet of the subsequent

166

glacial stage advanced over these lavas. The subglacial móberg ridge that caps the succession was built in an eruption beneath this ice sheet and it records the second glacial stage in the Búlandshöfði sequence.

Locality 9.2 Snæfellsjökull central volcano – a journey to the centre of the Earth
At the tip of the Snæfellsnes Peninsula is the perfectly cone-shape Snæ-fellsjökull stratovolcano, crowned by a summit caldera 3 km wide. The volcano is 1446 m high and on a clear day it is clearly visible from distant places such as Reykjavík. Because of its majestic stature, it was for long thought to be the highest mountain in Iceland (i.e. before accurate mapping). However, the volcano is perhaps best known for being the starting point of the epic journey portrayed by Jules Verne in his classic adventure story, *A voyage to the centre of the Earth.*

Snæfellsjökull volcano has been active for the past 700 000 years. In the past 10 000 years it has produced about 25 eruptions, all prehistoric. Three of those were Plinian eruptions, which occurred 9000, 3900 and 1750 years ago and dispersed rhyolitic tephra over parts of the Snæfellsnes Peninsula. The remainder were small basaltic fissure eruptions with lava and tephra emanating from parasitic vents at the foot of the volcano.

Locality 9.3 Eldborg á Mýrum – a spatter-cone volcano

That night an earth fire broke out and the Borgarhraun lava was formed. The farm stood where the cone is standing now.

This succinct description is all that was written in the Book of Settlement about the eruption that occurred in the Mýrar region in about 900. Initially, it was thought that this description referred to the formation of the fort-like spatter cone Eldborg (at Mýrar) and its lava flow. Later studies have shown that it is not so, because the eruption that formed Eldborg occurred more than 5000 years ago. However, the description is not entirely fictional because tephrochronology shows that the nearby Rauðhálsar cone and lava flow were formed in an eruption shortly after the first settlers arrived in Iceland.

The Eldborg lava is the largest (33 km^2) of six small lava flowfields that cover the lowlands of Mýrar and were formed by eruptions on short west-northwest trending fissures within the Ljósufjöll volcanic system. Eldborg is the largest cone of five that delineate a 1 km-long fissure. It is a complete spatter cone and it would be difficult to find a better example (Fig. 9.3). The cone is ellipsoidal in plan view, 250 m long and 180 m wide, and rises 50 m above its surroundings. The crater rims are steep (40–60°) and on the outside they are draped by small fountain-fed lava streams. During the

167

Figure 9.3 Aerial view of the spatter cone Eldborg at Mýrar.

eruption a small lava pond occupied the crater, and overflows from this pond formed the many 2–6 cm-thick lava units exposed in the inner crater walls. Most of the lava was transported away from the vents in subsurface lava tubes to form the bulk of the flowfield produced by the Eldborg eruption. However, the darker and less vegetated lava apron immediately around the cone is fountain-fed flows formed during the later stages of eruption.

Locality 9.4 The Borgarfjörður anticline
Travelling east on Highway 1 away from the town of Borgarnes and towards Bifröst, take note of the dip direction of the basalt-lava flows poking through the vegetation cover on either side of the road. North of the road the lava beds dip to the north-northwest, whereas on the south side they dip to the southeast. This reversal of dip direction across the road is explained by the fact that here it follows the axis of the Borgarfjörður anticline (see Fig. 9.1).

Locality 9.5 The Grábrók scoria cone volcano and the Hreðavatn unconformity
Grábrók is the largest cone on a 2 km-long scoria cone row. The eruption that formed the Grábrók scoria cone and the surrounding lava flow occurred 3600 years ago and represents the easternmost volcanic fissure in the Ljósufjöll volcanic system. Ready-made trails on the outer slopes of Grábrók make it quite suitable for a hike to the top, which provides an excellent view over the fertile regions of Borgarfjörður.

Nearby in the Brekkuá gorge is an outstanding exposure of the Hreðavatn sedimentary horizon and the associated unconformity. Here the 8 million-year-old Hreðavatn sediments consist of fossiliferous siltstone and sandstone beds, which rest unconformably on an eroded and southeast-dipping (10–15°) lava-flow sequence formed some 12–13 million years ago (Fig. 9.4). The lava sequence above the Hreðavatn horizon rests conformably on the sediments, and small pillow-like lobes occur at the base of the lowest lava, implying that it was emplaced into standing water.

On the way. Skorradalur, 2–3 km wide and 25 km long, is a classic example of a U-shape glacial valley. Lake Skorradalsvatn (15 km long, 43 m deep) sits in the glacially eroded depression behind a threshold of solid rock at the valley mouth. Initially, the valley was filled to the brim with glacial ice, which was reduced to a valley glacier as the climate became warmer during the waning stages of the Weichselian glaciation. As the glacier retreated farther about 12 000 years ago, the valley was inundated by the ocean, as is indicated by the fossil shells found in marine sediments at the valley mouth. Marine sediments of the same age are also found along the coast at Melabakkar to the west of Mt Hafnarfjall. There

6.5 million years old lava flow with pillows at the base

fossiliferous redeposited tuff

fossiliferous siltstone and sandstone

lava, 12–13 million years old, dipping southwest

Figure 9.4 Schematic cross section across the Hreðavatn unconformity at the Brekkuá gorge in Norðurárdalur, West Iceland.

the sediments are capped by tillite formed as end moraines in front of the Younger Dryas glacier. This sedimentary sequence indicates that the Borgarfjörður region was below sea level for more than 2000 years at the end of the Weichselian glaciation.

Mt Hvalfell is a small table mountain at the bottom of Hvalfjörður, formed in a subglacial eruption more than 15 000 years ago. It completely fills in the valley, such that, when the glacier retreated, a small lake was formed in the valley above the mountain (Fig. 9.5). Lake Hvalvatn is about 160 m deep, the second deepest lake in Iceland.

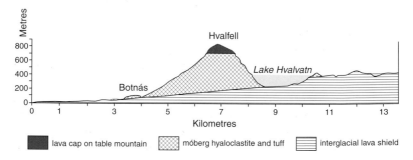

lava cap on table mountain

móberg hyaloclastite and tuff

interglacial lava shield

Figure 9.5 Cross section of Botnsdalur in Hvalfjörður, West Iceland. Lake Hvalvatn was formed when the table mountain Hvalfell was built by a subglacial eruption towards the end of the Weichselian glaciation, thus forming a natural dam across the valley.

170

Chapter 10

The Highlands

General considerations

We present the two highland excursions in a succession from north to south or vice versa; routes and stops are shown in Figures 5.1 and 7.1. Each of these excursions can also be undertaken as separate tours. For example, Askja–Kverkfjöll can be done individually with Lake Mývatn as a starting point and similarly the Landmannalaugar–Veiðivötn excursion can be undertaken from any of the major centres in South or Southwest Iceland. However, before delving into the descriptions, we would also like to offer a few words of advice. A trip across the Central Highlands in Iceland is not to be treated lightly and one should be prepared to face the elements in this harsh environment. Travel into the highlands without the guidance of experienced personnel is not recommended, other than for persons with serious mountaineering experience. There are no roads, only tracks, and only a few of the rivers have been bridged. Stock up with provisions before you leave, because there are no shops in the region.

The journey to Askja and the Kverkfjöll volcanoes can be accomplished in several ways. Some may do it the hard way, on foot or by mountain bike, whereas others can enjoy the luxury of a four-wheel-drive vehicle. The third option is to take the bus from the Hotel Reynihlíð at Mývatn, which runs daily over the summer time. The journey to Landmannalaugar and Veiðivötn can also be conducted in similar fashion. There are scheduled bus tours to the area from Reykjavík, Selfoss and Hella.

Askja–Kverkfjöll

On the way. Travelling from Mývatn towards the Hrossaborg tuff cone, where the turn-off to Askja is located, it is worth stopping at the top the Námafjall Ridge. There a magnificent view opens up towards the southeast over the northern part of Ódáðahraun, the largest wasteland of lava in Iceland. Above the endless lava fields rise many móberg ridges and

á house, æ **eye,** au feuille, ð **that,** ei day, é yet, hv quick, í seen, ö turn, ú soon, þ think table mountains, and the most prominent of these are the table mountains Búrfell and Bláfjall. The oldest lavas are more than 3000 years old and most of them are derived from the lava shield Ketildyngja. The Búrfell Lava, which covers the area immediately east of Námafjall, was erupted by the cone row Kræðuborgir and is about 2000 years old. The youngest lava is Nýja- hraun ("new lava"). It originated at the Sveinaborgir cone row in 1875, a 25 km-long volcanic fissure, known locally as Sveinagjá. The eruption of this lava occurred on 18 February 1875 during the Askja Fires, a volcanotectonic episode within the Askja volcanic system that began in 1874 and lasted until 1876. Here, out on the Askja fissure swarm, these fires were characterized by effusion of basaltic magma, whereas at the Askja central volcano the activity was very different. There the main phase of the activity was a Plinian eruption (see p. 175).

On the east side of Námafjall the highway traverses a gently undulating landscape, the product of north–south-trending graben formed by faulting in association with activity on the Krafla, Fremri–Námur and Askja fissure swarms. The characteristic feature of these graben systems is their limited width in relation to their length. For example, the previously mentioned 30 km-long Sveinar graben is nowhere more than 0.6 km wide. The vertical displacement on the graben faults is 2–20 m. Many of the graben systems contain one or more cone rows, and the age relations between the faults and the cones show that in many instances the graben were formed earlier than the cone rows.

About 3 km before the large bridge across Jökulsá á Fjöllum, turn onto the track leading towards the interior by a ring-shape volcanic cone, Hrossaborg, situated 1.2 km to the south of Highway 1. Road signs should indicate Herðubreiðarlindir and Askja. We are now well within the domain of the Askja volcanic system.

The Askja volcanic system – geology and history
The 200 km-long and less than 20 km-wide Askja volcanic system is the longest of its kind in Iceland (see Fig. 1.5). The central volcano, Askja (1550 m high), is situated in the southern sector of the system. The volcano has been active for several hundred thousand years. It rises to more than 800 m above its surroundings and is capped by an 8–10 km-wide composite caldera (see Table 1.1). The northern limb of the Askja fissure swarm is about 150 km long and contains many Holocene basalt volcanoes, includ ing the previously mentioned, Sveinagjá fissure and the Hrossaborg tuff cone. Farther south are the table mountain Herðubreið and the móberg ridge Herðubreiðartögl, which are surrounded by the lava shields Kollóttadyngja, Flatadyngja and Svartadyngja. The southern limb of the fissure swarm is much shorter at about 30 km long. The main Holocene basalt

eruptions are those of Gígöldur, Holuhraun and Þorvaldshraun. The latter two happened in historical times.

On the way. Hrossaborg ("horse pen") is a tuff cone formed by a explosive hydromagmatic eruption on the Askja volcanic system more than 7000 years ago. It features excellent outcrops for exploring the internal architecture of a tuff cone. The cone walls expose a range of hydromagmatic fall, surge and flow deposits, as well as impressive ballistic impact structures. Its irregular shape is attributed to collapse of the outer crater walls during the eruption, although it was aided by erosion from jökulhlaups that have occasionally rushed down the Jökulsá River bed in the past several thousand years.

Beyond Hrossaborg the geology is rather monotonous all the way to Herðubreiðarlindir, and the road follows an old sandur plain formed by torrential jökulhlaups induced by volcanic activity at or near the Kverkfjöll central volcano, some 100 km to the south. At Grímstaðanúpur the track lies across pahoehoe lava from the lava shield Kerlingardyngja, which was formed by a prolonged effusive eruption in the early Holocene. This is a good place to inspect pahoehoe lava with well developed tumuli and ropy lava surfaces.

Locality 10.1 Herðubreiðalindir – an oasis in rugged lava landscape
When arriving at Herðubreiðarlindir we first cross two rivers, Grafarlandaá and Lindá. Both are spring fed with clear mountain water. The springs are located at the margins of the lava flow in front of you to the west. These lavas belong to the 4000-year-old lava shield Flatadyngja ("flat shield") and act as a natural water filter. The rainwater seeps right into the porous rock, and impurities are filtered from it as the water is transported to the lava margins as groundwater. Thus, when the water emerges as springs it is not only crystal clear but also refreshingly chilled.

Herðubreiðarlindir ("oasis of Herðubreið"; 1682 m; Fig. 10.1), is one of the largest table mountains in the Central Highlands. This is an excellent place to spent the night, using the evening to explore the geology in the area. The lower part of Herðubreið is mainly built of subglacial volcanics, and the mountain is capped by a 300 m-thick lava shield. The transition from subglacial to subaerial volcanics occurs at about 1350 m, indicating that the ice sheet was at least 800 m thick when the subglacial eruption that sculptured the mountain into its present shape took place.

On the way (road to Drekagil at Askja). The first part of the route takes you through pahoehoe flows of the Flatadyngja lava shield, and then through rougher basalt flows that are derived from the Askja volcano. To the west of the track is the móberg ridge Herðubreiðartögl ("tail of Herðubreið"). Here you may notice a patchy cover of pale pumice on the

Figure 10.1 The majestic table mountain Herðubreið (background right) viewed from the south. In the foreground are older lavas and cones covered by the 1785 Askja Plinian fall deposit (whitish yellow), which in turn is partly covered by the 1961 Askja lava flow (black).

lava flows. These patches are the remnants of the rhyolite fall deposit from the 1875 Plinian eruption at Askja (see p. 175). Progressing south towards Askja volcano, the pumice fall covers a larger and larger area, until it completely blankets the surface at the southern end of Herðubreiðartögl and thickens rapidly towards the volcano.

The southern end of Herðubreiðartögl is a good stop to take a brief look at the surroundings. The Kverkfjöll central volcano rises above the horizon in the south, and the vast ice expanse in the background is the mighty Vatnajökull ice sheet. Looking to the west, the view opens to the Askja volcano and to the east the pillow lava ridges Miðfell and Upptyppingar rise above the lava plain. Still farther east is the móberg ridge Fagradalsfjall and, if weather permits, the ice-capped Snæfell stratovolcano is visible. Here the 1875 pumice rests on the early Holocene (>7000 years old) lavas of Svartadyngja ("black lava shield"), whereas a little farther south it rests on more rugged lavas that originated from vents within the Askja caldera about 8000 years ago. In front of Drekagil ("dragon's gorge") at the foot of the Askja volcano is a mountain hut, nicknamed Dreki ("dragon"), a rather rudimentary camping ground. Drekagil features excellent outcrops of products of subglacial eruption such as pillow lavas, kubbaberg breccia and móberg tuff.

Locality 10.2   The Askja volcano
From Dreki, the track meanders through a pristine basalt-lava flow, gradually climbing to Öskjuop, a 1 km-wide pass through the eastern wall of the Askja caldera, situated at an altitude of 1100 m. A pitch-black lava stands out against the pale pumice-fall deposit of 1875 that covers most of the older lava formations at Öskjuop (see Fig. 10.1). This is the youngest lava flow at Askja, which was formed in an effusive eruption in 1961.

The 1961 eruption began on 26 October on a 0.7 km-long fissure running east–west along the eastern margins of the main Askja caldera. In the beginning, the lava was spurting out as 500 m-high fountains, which within ten hours had produced a 7.5 km-long fountain-fed aa lava, which covered an area of 6 km². This was followed by an eruption of smooth-surface pahoehoe lava, which enlarged the lava flowfield to 11 km² by the time the eruption ended five weeks later. The spatter and scoria cones that delineate the 1961 fissure are the ones that rise over the parking lot at Öskjuop.

From the parking lot there is about a 20-minute walk to the explosion crater Víti and Öskjuvatn ("Askja lake"), which occupies a small collapse caldera formed in the Plinian eruption in 1875. The Öskjuvatn caldera is nested within the much larger main Askja caldera, which measures 8–10 km across. The steep-sided mountains and cliffs that surround this area on all sides are the rim of the main caldera.

The 1874–1929 Askja Fires   The 1874–1929 volcanotectonic episode, which includes the explosive Plinian eruption in March 1875, comprises the first recorded historical eruptions within the Askja volcanic system. This episode involved rifting along the Askja volcanic system and with eruptions at the central volcano and out on the fissure swarm. Therefore, this episode is collectively referred to as the Askja Fires.

Observations from the Mývatn area, some 100 km to the north, indicate that precursory activity began as early as February 1874, when dense columns of steam were seen rising from Askja. In the last two weeks of December 1874, strong and frequent earthquakes were felt in northern Iceland, and columns of smoke and fires were seen rising from the Askja region on 1 and 3 January 1875. Minor ashfall occurred in the districts south of Öxarfjörður in North Iceland. The Holuhraun basalt-lava flow to the south of Askja may have been formed in this phase of the Askja Fires.

A team of farmers from Mývatn visited the Askja volcano on 15 February 1875. They noticed that the southeast corner of the main caldera (i.e. the area of present-day Öskjuvatn) had subsided about 10 m and that a crater spewing mud and debris had formed just west of the subsidence. Three days later a basaltic fissure eruption began on the 30 km-long

175

Sveinagjá fissure some 40 km north of the Askja volcano. This eruption lasted for several months and produced 0.2–0.3 km^3 of lava.

An explosive Plinian eruption began within the Askja caldera in the early hours of 29 March. The eruption plume was dispersed eastwards by strong winds and by 3.30 a.m. tephra fall was reported in Jökuldalur about 70 km to the east of Askja. The tephra was greyish in colour and had a fine-grain texture like ash; it was also very wet and sticky. This first phase of the explosive eruption lasted for an hour. At 5.30 a.m. the tephra fall had dwindled and the air had become somewhat clearer. But this was just the calm before the storm, because shortly afterwards the main Plinian eruption began. Light-brown pumice of ever-increasing grain size rained down until noon. The plume was carried onwards across the Norwegian Sea, and tephra fall was reported in Scandinavia. Shortly after noon on 29 March the tephra fall had stopped over populated areas in Iceland, but the eruption raged on at Askja until the following day. Soon afterwards, on the north margins of the lake the maar-like explosion crater Víti was formed in a shortlived hydromagmatic eruption. Today, it contains fumaroles and a small lake with water warm enough for bathing.

Activity resumed in the Askja system between 1920 and 1930, with five separate eruptions occurring on ring fractures around the Öskjuvatn caldera and an eruption on a 6 km-long fissure on the southern flanks of the volcano that produced the Þorvaldshraun lavas.

In 1876 the Danish geologist Johnstrup visited the Askja volcano, together with his fellow countryman, the surveyor Caroc. They observed that a roughly round subsidence had formed in the southeastern corner of the Askja caldera. It measured 4580 m by 2500 m and the deepest part was 234 m below the floor of the main Askja caldera. This subsidence was bordered by many concentric faults and at its bottom was a small lake. Johnstrup and his companion were observing a caldera collapse. The roof of the crustal chamber beneath the Askja volcano had collapsed into the void left behind by the magma withdrawn from the holding chamber during the 1875 Plinian eruption. The Öskjuvatn caldera was born, the first of its kind in Iceland in historical times. Öskjuvatn did not reach its present water level until 1907; thus, it took 32 years to fill the depression with water.

The present level of the lake is 50 m below the caldera floor, and recent bathmetric soundings reveal a maximum water depth of 224 m. Areas of vigorous geothermal activity are found along the eastern and southern margins of the lake. Its area is 10.7 km^2 and the volume 1.2 km^3. The total volume of the subsidence is about 2 km^3, which is more than double the total volume of erupted magma. This suggests that a considerable amount of magma removed from the magma chamber during the 1875–1924 Askja Fires is stored as dyke intrusions within the volcano or the fissure swarm.

The Kverkfjöll volcanic system – geology and history
The Kverkfjöll volcanic system (see Fig. 1.5) is about 125 km long. The main volcanic structures are the Kverkfjöll central volcano (1929 m high) with its glacier-filled twin calderas and the mountain range Kverkhnjúkar, which constitute the most volcanically active part of the fissure swarm. One of the outstanding features of Kverkhnjúkar is a series of volcanic ridges, 10–30 km long, that extends northwards from the central volcano and rise 200–300 m above the surrounding plains (Fig. 10.2). These ridges are composed almost entirely of pillow lava, which implies that they were formed by effusive fissure eruptions under a substantial thickness of glacial ice. Tuffs formed by explosive subglacial eruptions occur only on peaks that rise to more than 1000 m above sea level, showing that the Weichselian glacier was at least 400 m thick when these subglacial eruptions took place.

The valleys between the ridges are floored by many basalt-lava flows formed by subaerial fissure eruptions over the past 10 000 years. Some of these eruptions must have featured spectacular lava fountains, because they produced extensive aprons of fountain-fed lava mantling every hill and peak up to 1.5 km from the source vents.

Locality 10.3 Hvannalindir – a young oasis in a highland desert
Hvannalindir is a small oasis in the otherwise barren highlands north of the Kverkfjöll volcano. It is situated in front of the Lindahraun lava and owes its existence to the freshwater stream Lindá that emerges as cold-water springs at the lava front on the south side of Hvannalindir (Fig. 10.3a).

The sandy soil cover at Hvannalindir is up to 2 m thick and contains several tephra layers and three thick units of aeolian sand. The oldest tephra layer in the soils of Hvannalindir indicates that the oasis began to form about 1300 years ago, two centuries before the human colonization of Iceland (Fig. 10.3b). More importantly, this also gives the minimum age (AD 700) for the Lindahraun lava, which is the youngest volcanic formation on the Kverkfjöll fissure swarm. These results show that no volcanic eruption has occurred on the Kverkfjöll volcanic system outside of the central volcano in historical times (i.e. the past 1200 years), which is surprising because the Kverkfjöll system is situated above the centre of the mantle plume.

The youngest tephra layer in the soil at Hvannalindir is the rhyolite tephra fall from the 1362 eruption at Öræfajökull. This shows that, until the fourteenth century, the climate was favourable and the Hvannalindir oasis enjoyed stable and perhaps lush vegetation cover. The aeolian sand unit towards the top of the soil cover shows that some time after the

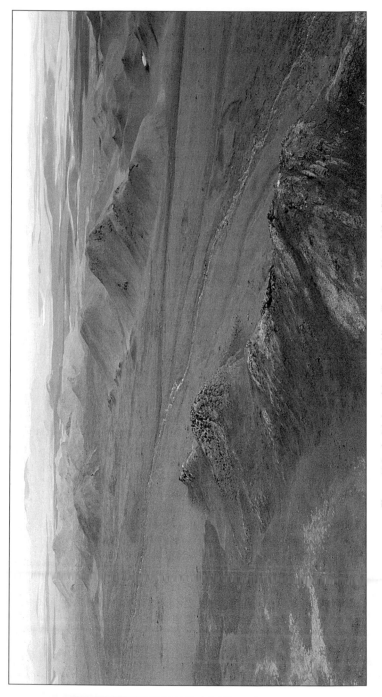

Figure 10.2 Aerial view of the Kverkhnjúkar pillow-lava ridges at Kverkfjöll.

fourteenth century the conditions changed for the worse. The climate deteriorated as the Little Ice Age set in and, because of stronger winds, the oasis was inundated by sand instead of silty loess; thus, the vegetation deteriorated. The oasis has not yet recovered from the setback of the Little Ice Age.

In the eighteenth, century Hvannalindir was the refuge of Iceland's most famous outlaw, Fjalla-Eyvindur ("Eyvindur of the mountains"). In those days it was a perfect hiding place because it was secluded and protected by large glacial rivers on three sides and on the fourth by the Vatnajökull glacier. Fjalla-Eyvindur lived here for almost a decade. The ruins of his home can be seen near the edge of the Lindahraun lava in the southeast corner of the oasis. Here Fjalla-Eyvindur built a rather elaborate home out of rocks and turf, containing separate sleeping quarters, a kitchen, and a toilet that included an automatic flushing system (the river), which would have been a luxury in those days.

Landmunnalaugar–Veiðivötn

This excursion is centred on the popular Torfajökull–Veiðivötn region, which features some of the most colourful rock formations in Iceland and spectacular rows of explosion cones and craters.

Locality 10.4 The Torfajökull volcano – complex accumulation of rhyolite lavas and domes

The Torfajökull central volcano in South Iceland is the largest caldera volcano in Iceland, with a basal diameter of 30 km, and is crowned by a 18 km-wide caldera (see Table 1.4), which hosts the largest geothermal area in Iceland. This rugged mountain region is by far the largest area of rhyolite in the country, where colourful subglacial and subaerial volcanics cover about 350 km^2 (Fig. 10.4). The Torfajökull volcano is an anomaly in the predominantly basaltic volcanic succession of Iceland, because the ratio of rhyolites to basalt is about four to one, which is in stark contrast to the 1:5 ratio observed for most of the other central volcanoes. This anomaly reflects the setting of the volcano, which is situated at the boundary where the East Volcanic Zone changes from being a simple rift zone, characterized by spreading away from the rift axis, to a forward-propagating rift zone. The Torfajökull volcano exhibits distinct polarity in terms of the composition of the erupted magmas. It produces mostly rhyolites, and basalts of intermediate (andesitic) compositions are present only in minor amounts. This compositional gap, often referred to as the Daly gap, implies that the basalt and rhyolite magmas are not derived from the same

179

Figure 10.3 **(a)** Geology of the area around Hvannalindir, showing the distribution of the Lindahraun lava and other geological formations. **(b)** Tephra layers in a soil profile at Hvannalindir. The numbers are years of eruption, with the source volcanoes named.

source. The basalt magmas originate from a deeper source within the mantle plume, whereas the rhyolite magmas are produced by partial melting of basaltic crust at shallow depths (3–10 km).

The oldest stratigraphical units at the Torfajökull volcano date back to the most recent interglacial period (100 000 years ago). These are thick rhyolitic lavas and welded airfall tuffs. This was followed by large ring-fracture eruptions forming a series of subglacial rhyolite lava domes (e.g. Bláhnjúkur, Kirkjufell, Torfajökull, Laufafell and Rauðufossafjöll) that encircle the entire caldera (Fig. 10.4). Some of these lava domes grew large enough to penetrate the ice sheet and form silicic table mountains.

In postglacial times the Torfajökull volcano has produced several relatively thin (< 50 m) obsidian lava flows that appear as dribbles of black icing on the brightly coloured landscape. These obsidian lavas are Dómadalshraun, Hrafntinnuhraun ("obsidian lava"), Námshraun and Laugahraun (Fig. 10.4). Dómadalshraun is the oldest, formed in AD 150, then Hrafntinnuhraun (870), followed by the eruption of Námshraun and Laugahraun in 1480. All of these lavas were erupted in association with major basalt-fissure eruptions on the adjacent Veiðivötn volcanic system (see locality 10.5) and they also contain an abundance of mafic-melt inclusions,

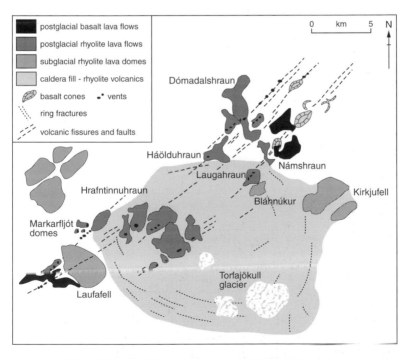

Figure 10.4 Simplified geological map of the Torfajökull volcano.

indicating that rhyolite residing in a shallow crustal magma chamber was reheated and mobilized by injection of hotter basalt magma. The youngest rhyolite formations at Torfajökull are erupted on northeast–southwest orientated fissures, implying that at present the activity at the volcano is mainly controlled by normal rift-zone tectonics.

Locality 10.5 Veiðivötn and Vatnaöldur fissures, cones and maars
The Veiðivötn volcanic system (see Fig. 1.5) is about 150 km long and takes its name from the many lakes and ponds of Veiðivötn ("lakes of good fishing"), which are known for their trout. The most prominent volcanic landforms in the system are the glacier-covered Bárðarbunga central volcano at the western edge of Vatnajökull icecap and the volcanic fissures that transect the Veiðivötn basin from northeast to southwest. The Veiðivötn system is one of the most productive volcanic areas in Iceland in Holocene and historical times. In the past 800 years the Bárðarbunga volcano has erupted at least 17 times and at least 12 major basalt-fissure eruptions have occurred in the Veiðivötn basin in the past 10000 years (Table 10.1). The lavas produced by the Holocene flood-lava eruptions, which include the Great Þjórsá and the Þjórsárdalur flows (see pp. 71–72), generally followed the course of the Tungnaá and Þjórsá rivers as they found their way towards and onto the Southern Lowlands). This is a major concern, as lava flows from future eruptions may follow the same general path and destroy the hydroelectric powerplants on these rivers at and above Mt Búrfell.

Today, most of the Holocene fissures and craters are hidden from view by the many lakes situated in tuff cones, tuff rings and maar craters that floor the Veiðivötn basin. Most of these cones and craters were formed in three fissure eruptions in AD 150, 870 and 1480. The first of these eruptions occurred on an 8 km-long volcanic fissure at the southwestern end of the basin (Fig. 10.5). The southern end of the fissure extended into the

Table 10.1 List of Holocene and historical basaltic flood lava and fissure eruptions in the Veiðivötn area. Ages are listed as years before present, using 2000 as the reference year. Numbers in parentheses indicate calendar years (AD).

Lava flow or eruption	Age (BP)	Length (km)	Area (km²)	Volume (km³)
Great Þjórsárhraun	8600	130	950	22.0
Tungnaárhraun (THb)	6800	55	120	1.4
Tungnaárhraun (THd)	6700	80	270	3.8
Tungnaárhraun (THe)	6600	75	260	1.0
Sigolduhraun	6500	65	200	3.5
Hnubbahraun	6000	15	25	0.4
Búrfellshraun–Þjórsárdalshraun	3200	70	350	6.5
Tjörvahraun–Hnausahraun	1850 (150)	20	55	0.8
Vatnaöldur (explosive)	1130 (870)	–	–	3.3
Veiðivötn (explosive)	520 (1480)	–	–	3.5
Tröllahraun lava	138 (1862)	25	28	0.3

LANDMANNALAUGAR–VEIÐIVÖTN

Torfajökull volcano and produced the silicic Dómadalshraun (see p. 179), whereas the northern part of the fissure-erupted basaltic magma and formed two small basaltic lava flows known as Hnausahraun and Tjörvahraun. These lavas flowed into and dammed up the Tungnaá River, which consequently flooded the Veiðivötn area, forming a basin-wide lake in the process. Thus, when the 870 eruption took place and formed the Vatnaöldur cone and crater row, the basin was submerged in water (Fig. 10.5). As the magma emerged from the new fissures, it encountered and interacted explosively with the lake water. As a result, the eruption was almost entirely explosive and it covered roughly half of Iceland with basaltic tephra. As in the previous eruption, the feeder dyke for the Vatnaöldur eruption cut through the Torfajökull volcano and thus initiated an eruption at Hrafntinnusker, which produced rhyolite tephra, along with the obsidian lava of Hrafntinnuhraun. The Vatnaöldur eruption pre-dates the arrival of the first settlers in Iceland by a few years, and its tephra layer is also known as the Settlement layer, which is an important time marker in the soils of Iceland for both volcanological and archaeological studies. After the Vatnaöldur eruption, the lake was re-established and thus the events of the 870 were repeated when the 1480 eruption occurred and produced the 40 km-long Veiðivötn cone and crater row (Fig. 10.5). The Vatnaöldur and Veiðivötn eruptions produced 3.3 km^3 and 3.5 km^3 of basaltic tephra, respectively, which makes them the two largest hydromagmatic basalt fissure eruptions in recent history.

Figure 10.5 The main features and position of the AD 150 Hnausahraun–Tjörvahraun, 870 Vatnaöldur, and 1480 Veiðivötn vent systems. The broken line shows the approximate outlines of the Langisjór lake formed as a consequence of the eruption in AD 150.

Glossary

bedding plane   The plane separating sediment beds.

caldera   A large (>1 km in diameter) circular depression formed by downsagging or collapse of a magma chamber roof as a result of withdrawal of magma from the chamber during eruption(s).

compound lava   Lava flows that consist of multiple flow lobes in various lateral and vertical arrangements. Note that almost all lava flows consist of two or more flow lobes and thus in essence are compound. However, lavas that comprise long and thick flow lobes may superficially appear to consist of a single flow lobe (or flow unit). Such lavas are sometimes referred to as "simple flows".

corrie or **cirque**   Landform of glaciated highland regions, created when there is enough snow accumulation at the head of an ordinary valley head to form a glacier large enough to begin to flow. The backwalls of the valley are cut back by plucking, whereas the valley floor is gouged out by the moving glacier. This erosion by the glacier forms a bowl-shape depression called a corrie or a cirque.

debris flow   Rapid downslope movement of a mass of debris that exhibits the flow properties of a viscous fluid.

diamictite   A non-genetic term used for chaotic and unstratified sediments that consist of clay- to boulder-size particles and are poorly sorted.

diatom   A one-cell plant that has siliceous framework and grows in oceans and lakes.

differentiation (magma evolution)   Magma has the tendency to rise towards the surface because it is less dense than the surrounding material.

In doing so it moves to cooler environments and begins to crystallize. In most natural systems, the composition of the magma changes during the travel from the point of origin to the location of solidification, because magmas crystallize over a large temperature interval. The main reason for this behaviour is that, upon cooling, the first minerals to crystallize are those with the highest freezing (or melting) points, whereas those with the lowest will be the last minerals to form. Thus, early on in the process the magma consists of two **phases**, melt and a mineral. If the early-formed minerals (e.g. olivine, Mg_2SiO_4) are removed from the magma during transport by some mechanism, then the bulk composition of the magma will change. The process has removed Mg and SiO_4 from the melt in the ratio of 2:1. Consequently, twice as much Mg as SiO_4 is removed from the magma, whereas concentration of other elements such as Al, Na, and K is unchanged. The removal of the olivine minerals has also reduced the bulk volume of the magma body. When corrected for this volume change, it is found that the olivine-free magma has been depleted in Mg and, relatively speaking, enriched in all other components, including silica. Thus, the magma is said to have **evolved** to more silica-rich composition through the process of **differentiation**. Such differentiation characterizes many volcanoes, which often begin by erupting mafic (parental) magmas and end their life cycle by producing mainly felsic (daughter) magmas. Mixing of different magmas can also change the bulk composition and is another important process for magma evolution.

dropstone   Ice-rafted gravel to boulder size stones (i.e. rock fragments),

185

which drop from melting iceberg into marine sediments.

eutaxitic texture   See **fiammé**.

fault   A fracture along which the opposite sides have been displaced relative to each other

faulting   Deformation of crustal rocks by fracture.

fiammé   Deformed pumice clasts that occur in welded ignimbrites as pale brown to black glassy lenses with flame-like shapes. Fiammé exhibit bedding-parallel alignment of long dimensions, giving the rock a distinctive foliation referred to as **eutaxitic texture**.

fires   In the old chronicles, Eldur ("fire") is often used to describe a volcanic eruption. It reflects direct experience of eruptions, which most commonly feature fire-fountain activity or, as perceived, columns of fires.

fold   A pronounced bend in layers of rock.

folding   Deformation of crustal rocks by bending.

graben   A downthrown block between two parallel normal faults.

hornito   A small spatter (i.e. driblet) cone formed around rootless vent on a lava flow.

horst   An elongate uplifted block bounded by faults on its long side.

hotspot   Area of unusually productive volcanism generally considered to be the surface manifestation of a mantle plume.

hyaloclastite   Volcaniclastic breccia formed by non-explosive fracturing and disintegration of quenched lavas. Fragmentation occurs in response to thermal stress built up during rapid cooling and stress imposed on chilled outer parts of lava flows by continued movement of the molten interior.

intraplate volcanic belts   A volcanic zone located within a plate rather than at the boundary between two plates.

isochron   A line connecting points of equal age.

jökulhlaup   An Icelandic term for a glacier outburst flood.

lava fountain   A red-glowing pillar or column of lava rising into the air from an active vent.

large igneous province   A very large basaltic volcanic province characterized by emplacement of mainly mafic (i.e. Fe- and Mg-rich) extrusive and intrusive rocks, which originate via mantle processes other than normal seafloor spreading. For example, they are produced by mantle plumes.

lithification   The process that changes a sediment into sedimentary rock

magma   Molten rock that solidifies upon cooling to form igneous rocks.

mantle plume   Plume-shape structure of hotter and less dense material within the Earth's mantle that is more buoyant than the surrounding mantle and thus rises towards the surface.

mid-ocean ridge   A major elevated linear feature of the sea floor, marking the boundary between diverging plates. It typically features a series of 200–20 000 km-long segments offset by transform faults.

mineral   A naturally occurring solid inorganic element or compound that exhibits a regular internal crystal structure and a distinctive chemical composition. Studies have shown that more than 2000 minerals exist, but about 20 minerals are so abundant that they account for more than 95% of the mineral mass in the Earth's crust. These are known as the **rock-forming minerals**. The most common rock-forming minerals are silicates, named so because their chief constituents are oxygen (O) and silicon (Si), which bond to form silica (SiO_4). In turn, the silica bonds with one or more of the elements magnesium (Mg), iron (Fe), calcium (Ca), aluminium (Al), sodium (Na), and potassium (K) to form silicate minerals, such as olivine, pyroxene, amphibole, biotite, plagioclase, alkali feldspar, and muscovite.

obsidian   Dark volcanic glass of felsic composition.

oceanic crust   The crust underlying

the oceanic basins, mainly composed of basaltic volcanic rocks.

ogives A dark curved arcuate structure, one of a series repeated periodically down a glacier, typically formed at a base of an icefall, and resembling the pointed arch or rib across a Gothic vault.

palagonite An altered volcanic glass, brown to yellow or orange, commonly found on the outer margins of pillow lava lobes and in hydromagmatic tuffs.

partial melting Melting points of igneous minerals change systematically, where olivine and calcium-rich plagioclase have the highest melting temperature and quartz has the lowest. When a rock is heated, the minerals with low melting temperatures dissolve first and those with the highest melting temperatures are most resistant to melting. If all of the minerals that make up a rock are melted, then the magma will have the same bulk composition as the rock. On the other hand, if only the minerals with the lowest melting point are dissolved in the process, then the rock has been subjected to partial melting and will have different composition than the rock from which it is derived. At relatively low temperatures (700–900°C) only the felsic minerals, such as quartz, feldspar, or muscovite are dissolved, and melting produces **felsic** magmas. Conversely, melting at high temperatures (>1100°C) yields melts that are rich in olivine, plagioclase and pyroxene, and thus form **mafic** magmas. Melting at temperatures between these two extremes may form magmas of intermediate compositions.

pyroclast An individual fragment or clast in pyroclastic deposits or rocks formed by disintegration of magma in explosive eruptions.

regolith Any solid material lying on top of bedrock, including soils, alluvium, and rock fragments weathered from the bedrock.

rise velocity The speed at which the material rises.

sandur An Icelandic word, it is used internationally to denote broad plains of glaciofluvial deposits formed by braided river systems.

spherulite Radiating aggregates of alkali feldspar, 0.1–2 cm in diameter, formed by crystallization of volcanic glass while the lava is still hot, but at temperatures below the freezing point of the magma. This process is known as devitrification.

stony rhyolite Crystalline felsic lava formed by devitrification of volcanic glass during slow cooling. Typically features fine-grain intergrowths of quartz and alkali feldspar known as felsitic texture.

strandlines An ephemeral line or level at which a body of water (i.e. the sea or a lake) meets the land. Also commonly used to refer to shorelines now elevated above present water level

transform or **transform fault** A strike-slip fault connecting the ends of offset segments in mid-ocean ridges.

volcanic succession Volcanic rock units or mass of volcanic strata that succeed one another in chronological order.

volcaniclastic deposits A debris or consolidated clastic rock predominantly composed of volcanic particles.

volcanotectonic Geological structures or events that result from or are controlled by both volcanic and tectonic processes.

Bibliography

Bergh, S. G. & G. E. Sigvaldason 1991. Pleistocene mass-flow deposits of basaltic hyaloclastite on a shallow submarine shelf, South Iceland. *Bulletin of Volcanology* **53**, 597–611.

Bjornsson, H. & P. Einarsson 1990. Volcanoes beneath Vatnajökull, Iceland: Evidence from radio-echo-sounding, earthquakes, and jokulhlaups. *Jökull* **40**, 147–67.

Bjornsson, A., K. Saemundsson, P. Einarsson, E. Tryggvason, K. Gronvold 1977. Current rifting episode in North Iceland. *Nature* **266**(5600), 318–23.

Blake, D. H. 1966. The net-veined complex of the Austurhorn intrusion, Southeast Iceland. *Journal of Geology* **74**, 891–907.

Einarsson, P. & S. Bjornsson 1979. Earthquakes in Iceland. *Jökull* **29**, 37–43.

Einarsson, Þ. 1994. *Geology of Iceland – rocks and landscape*. Reykjavík: Mál og Menning.

Geirsdóttir, Á. & J. Eiríksson 1994a. Growth of an intermittent ice sheet in Iceland during the late Pliocene and early Pleistocene. *Quaternary Research* **42**, 115–30.

———— 1994b. Sedimentary facies and environmental history of the late-glacial glaciomarine Fossvogur sediments in Reykjavik, Iceland. *Boreas* **23**, 164–76.

Gudmundsson, A. 1995. Infrastructure and mechanics of volcanic systems in Iceland. *Journal of Volcanology and Geothermal Research* **64**, 1–22.

———— 2000. Dynamics of volcanic systems in Iceland: example of tectonism and volcanism at juxtaposed hotspot and mid-ocean ridge systems. *Annual Review of Earth and Planetary Sciences* **28**, 107–140.

Gunnarsson, B. 1987. *Petrology and petrogenesis of silicic and intermediate lavas on a propagating oceanic rift: the Torfajökull and Hekla central volcanoes, South-Central Iceland.* PhD thesis, Johns Hopkins University, Baltimore.

Hjartarson, Á. 1991. A revised model of Weichselian deglaciation in South and Southwest Iceland. In *Environmental change in Iceland B past and present*, J. K. Maizels & C. Caseldine (eds), 67–77. Amsterdam: Kluwer.

Jakobsson, S.P., 1968. The geology and petrology of the Vestmann Islands. A preliminary report. *Surtsey Research Progress Report*, **4**, 113–29.

———— 1979. Petrology of recent basalts of the Eastern Volcanic Zone, Iceland. *Acta Naturalia Islandica* **26**, 1–103.

Jóhannesson, H. 1980. Jarðlagaskipan og þróun rekbelta á Vesturlandi. *Náttúrufræðingurinn* **50**, 13–31.

Jóhannesson, H. & K. Sæmundsson 1998. *Geological map of Iceland* (2nd edn). Reykjavík: Icelandic Institute of Natural History.

Jones, J. G. 1969. Intraglacial volcanoes of the Laugarvatn region, Southwest Iceland, 1. *Geological Society of London, Quarterly Journal* **124**, 197–211.

Jones, J. G. 1970. Intraglacial volcanoes of the Laugarvatn region, Southwest Iceland,

2. *Journal of Geology* **78**(2), 127–40.

Jónsson, J. 1988. Hestgcrðismúli í Suðursveit. *Náttúrufræðingurinn* **58**, 87–96.

Jökull (special issue on the geology of Iceland), **29**.

Kjartansson, G. 1973. Aldur Búrfellshrauns við Hafnarfjörð. *Náttúrufræðingurinn* **42**, 159–83.

Larsen, G. 1984. Recent volcanic history of the Veidivotn fissure swarm, southern Iceland: an approach to volcanic risk assessment. *Journal of Volcanology and Geothermal Research* **22**, 33–58.

——— 2000. Holocene eruptions within the Katla volcanic system, South Iceland: characteristics and environmental impact. *Jökull* **49**, 1–28.

Larsen, G. & S. Thorarinsson 1977. H4 and other acid Hekla tephra layers. *Jökull* **27**, 28–46.

Larsen, G., A. Dugmore, A. Newton 1999. Geochemistry of historical-age silicic tephras in Iceland. *Holocene* **9**, 463–71.

Larsen, G., M. T. Gudmundsson, H. Bjornsson 1998. Eight centuries of periodic volcanism at the center of the Iceland hotspot revealed by glacier tephrostratigraphy. *Geology* **26**, 943–6.

Lockwood, J. P. & P. W. Lipman 1980. Recovery of datable charcoal beneath young lavas. lessons from Hawaii. *Bulletin Volcanology* **43**, 609–615.

McPhie, J., M. Doyle, R. Allen 1993. *Volcanic textures: a guide to the interpretation of textures in volcanic rocks*. Hobart: Tasmanian Government Printing Office.

Norddahl, H. & H. Haflidason 1992. The Skógar tephra, a Younger Dryas marker in North Iceland. *Boreas* **21**, 23–41.

Oskarsson, N., G. E. Sigvaldason, S. Steinþorsson 1982. A dynamic model of rift zone petrogenesis and the regional petrology of Iceland. *Journal of Petrology* **23**(1), 28–74.

Sanders, A. D., J. G. Fitton, A. D. Kerr, M. J. Norry, R. W. Kent 1997. North Atlantic Province. In *Large igneous provinces: continental, oceanic, and planetary flood volcanism*, J. J. Mahoney & M. F. Coffin (eds), 44–93. Geophysical Monograph 100, American Geophysical Union, Washington DC.

Sigmarsson, O., C. Hemond, M. Condomines, S. Fourcade, N. Oskarsson 1991. Origin of silicic magma in Iceland revealed by Th isotopes. *Geology* **19**, 621–4.

Sigurgeirsson, M. Á. 1995. Yngra Stampagosið á Reykjanesi. *Náttúrufræðingurinn* **64**, 211–30.

Steinþórsson, S. 1966. The ankaramites of Hvammsmúli, Eyjafjöll, southern Iceland. *Acta Naturalia Islandica* **2**, 1–32.

——— 1981. Ísland og Flekakenningin. In *Náttúra Íslands*, 29–63. Reykjavík: Almenna Bókafélagið.

Steinþórsson, S., N. Oskarsson, G. E. Sigvaldason 1985. Origin of alkali basalts in Iceland: a plate tectonic model. *Journal of Geophysical Research* **90**(B12), 10027–10042.

Sæmundsson, K. 1979. Outline of geology of Iceland. *Jökull* **29**, 7–28.

——— 1991. Jarðfræði Kröflukerfisins. In *Náttúra Mývatns*, 25–95. Reykjavík: Hið Íslenska Nátturufræðifélag.

——— 1992. Geology of Thingvallavatn area. *Oikos* **64**, 40–68.

Thorarinsson, S. 1958. The Öraefajökull eruption of 1362. *Acta Naturalia Islandica* **2**, 1–100.

———— 1967. *The eruptions of Hekla in historical times: a tephrochronological study* [in the series The Eruption of Hekla 1947–48]. Special Publications 1, Societas Scientiarum Islandica, Reykjavík.

———— 1969. *Surtsey: the new island in the North Atlantic*. London: Cassell.

Thordarson, T. 2000. Morphology and structure of the AD 1963 –67 Surtsey lava flow field: results from preliminary observations. *Surtsey Research Program Report* **XI**, 125–42.

Thordarson, T. & S. Self 1993. *The Laki (Skaftár Fires) and Grímsvötn eruptions in 1783– 1785. Bulletin of Volcanology* **55**, 233–63.

———— 1998. The Roza Member, Columbia River Basalt Group B a gigantic pahoe- hoe lava flow field formed by endogenous processes. *Journal of Geophysical Research* **103**, 27411–45.

———— 2001. Real-time observations of the Laki sulfuric aerosol cloud in Europe 1783 as documented by Professor S. P. van Swinden at Franeker, Holland. *Jökull* **50**, 65–72.

Thordarson, T., D. J. Miller, G. Larsen, S. Self, H. Sigurdsson 2001. New estimates of sulfur degassing and atmospheric mass-loading by the AD 934 Eldgjá eruption, Iceland. *Journal of Volcanology and Geothermal Research* **108**(1–4), 33–54.

Walker, G. P. L. 1959. Geology of the Reydarfjordur area, eastern Iceland. *Quarterly Journal of Geological Society of London* **114**, 367–91.

———— 1960. Zeolite zones and dike distribution in relation to the structure of the basalts of eastern Iceland. *Journal of Geology* **68**, 515–27.

———— 1962. Tertiary welded tuffs in Iceland. *Geological Society of London, Quarterly Journal* **118**, 275–93.

———— 1963. The Breiðdalur central volcano, eastern Iceland. *Geological Society of London, Quarterly Journal* **119**, 29–63.

———— 1964. Geological investigations in eastern Iceland. *Bulletin of Volcanology* **27**, 1–15.

———— 1966. Acid volcanic rocks in Iceland. *Bulletin of Volcanology* **29**, 375–402.

Index

Following Icelandic convention, the ð/Ð (edh or eth) character falls between d and e; the þ/Þ character (thorn) follows z. Vowels follow conventional rules of alphabetical sequencing.